TOWER HAMLETS
91 000 001 912
D0586354

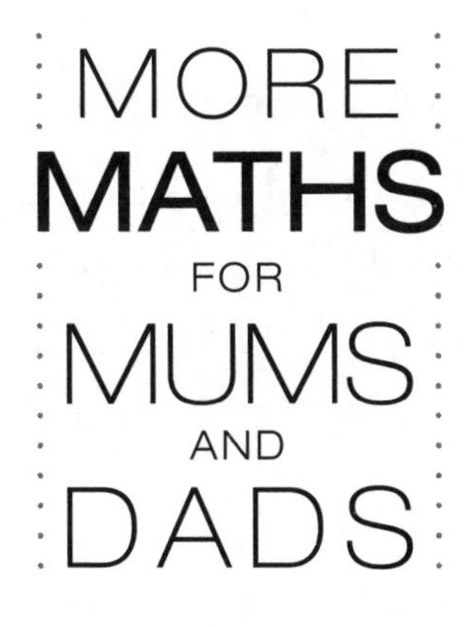
MORE
MATHS
FOR
MUMS
AND
DADS

Also by Rob Eastaway and Mike Askew

Maths for Mums and Dads

MORE **MATHS** FOR MUMS AND DADS

Rob Eastaway and Mike Askew

SQUARE PEG

Published by Square Peg 2013

2 4 6 8 10 9 7 5 3 1

First published in Great Britain in 2013 by
Square Peg
Random House, 20 Vauxhall Bridge Road,
London SW1V 2SA

www.vintage-books.co.uk

Addresses for companies within The Random House Group Limited can be found at: www.randomhouse.co.uk/offices.htm

The Random House Group Limited Reg. No. 954009

A CIP catalogue record for this book is available from the British Library

ISBN 9780224095310

The Random House Group Limited supports the Forest Stewardship Council® (FSC®), the leading international forest-certification organisation. Our books carrying the FSC label are printed on FSC®-certified paper. FSC is the only forest-certification scheme supported by the leading environmental organisations, including Greenpeace. Our paper procurement policy can be found at www.randomhouse.co.uk/environment

Printed and bound in Great Britain by the MPG Books Group

CONTENTS

PART ONE
MATHS, PARENTS & TEENAGERS

PART TWO
THE EVERYDAY SKILLS

PART THREE
THE MATHS

ALGEBRA

GEOMETRY

NUMBERS, CALCULATION & MEASUREMENT

PROBABILITY AND STATISTICS

PART FOUR
THE QUESTIONS & ANSWERS

INTRODUCTION

Your child has made it through primary school. While they were there, you probably discovered the new-fangled methods for arithmetic, such as chunking and the grid method; you might even have discovered that some of these modern methods are easier to understand than the traditional ones.*

But when your child embarks on maths at secondary school, two new issues arise. First, in the build-up to GCSE, school children begin to do maths that you have never encountered before – or if you have, you have long since forgotten it. Factorising equations? Finding the locus? Solving for x? What do these even *mean*?

And the second problem? As your child becomes a teenager, two dreaded questions increasingly loom: *When will I ever need this?* And even worse: *Who cares?*

And since the standard answer to these questions is often *You will find this useful one day if you go on to be such and such* – (response: *But I'm not going to be a such and such*) – things can become extremely fraught.

Also of course your teenager might well say that they don't actually *want* your help. This is partly because you are their parent and they want as little interaction with you as possible; but also perhaps because they don't believe that you *can* help, either

* If these somehow passed you by, the methods and their rationale are covered in *Maths for Mums and Dads*.

because they think you won't understand or because you're only going to confuse them by showing them different methods from the ones they do at school. (Both of these might be true.)

This book will help you with both problems – dealing with your child's attitude towards maths, as well as getting to grips with the maths yourself. We explain in straightforward terms the maths your child will encounter in their early teens. In many cases we give examples of where maths crops up in the real world. and for those topics that are more abstract we'll suggest ways to find some meaning in the maths. And we'll suggest strategies to make your teenager more likely to engage with maths – and engage with you.

(By the way, we're aware that if you happen to leave this book lying around at home, your teenager might get curious and want to dip into it. This mightn't be a bad strategy.)

A word of warning, however: this is NOT a textbook. The secondary curriculum is very broad, different schools teach the same topics in different ways, and it isn't possible to cover every single angle (excuse the pun) in this book. But if you want to know what is the *point* of, say, algebra and why a teenager should even bother to grapple with it in the first place, this book might be a better place to start.

PART ONE

MATHS, PARENTS & TEENAGERS

Teenager: Dad, can you do my maths homework for me?

Dad: No, it wouldn't be right.

Teenager: Well, why not have a go anyway? You might do better than you think.

WHAT TEENAGERS SAY

My best advice for parents would be: Leave me alone. Unless I ask for help.*

Most of the parents we talked to want to help their teenagers. Unfortunately, in talking to teenagers we found that a lot of them didn't want parental help. It's easy to assume that this is because many teenagers want to have as little contact with their parents as possible, but we discovered a variety of reasons why some would rather their parents leave them alone when it comes to maths. Two common reasons were summed up by these quotes from teenagers:

They just confuse me because they try to teach me ways that are different from the ones we learn at school.

They don't understand it themselves and so I have to wait around for ages while they try to work it out.

Another reason why teenagers avoid parents was nicely hinted at by this frustrated dad:

Typically, my children want me to just quickly tell them 'the answer', whereas I want to explore the question, make sure they

* Fourteen-year-old boy

understand ideas, can demonstrate the method and why the answer is as it is.

From a teenager's point of view this often comes across rather differently.

I like asking Mum because she just gives me the answer, whereas Dad always wants to give me a lecture.

On the other hand, many teenagers get frustrated by maths precisely because they do want to understand the bigger picture but get told instead not to worry about that and just learn a technique. So when your teenager does come to you for help, it's worth asking what sort of help they want – the quick fix or a chance to discuss what it's about.

Is there a different way to deal with boys and girls? There are two common stereotypes, that boys just want the quick answer, while girls need to understand and want to be careful and tidy. However, the reality is that there is as much variation within genders as between them. When we asked fourteen-year-olds to tell us the difference between the way boys and girls do maths, the most common answer was that 'there is no difference', and that it depends on the individual. You'll know what motivates your teenager.

'Fast' and 'Slow' Thinking – That Sinking Feeling

Most parents are only too aware that teenagers confronted with maths homework can get particularly grumpy. What is it about maths that makes it so prone to 'that sinking feeling' (as one teenager described it)?

There's a particular question that gives a clue. Have a go at it:

THE BOW AND ARROW QUESTION

A bow and an arrow cost £11 in total. The bow costs £10 more than the arrow. What does the arrow cost?

You probably came up with an answer to this question almost instantly. Teenagers certainly do.

If your answer is £1, then you agree with the vast majority of teenagers and adults.

But if £1 is your answer, just think for a second. The bow costs £10 more than the arrow, so if the arrow cost £1 then the bow cost £11, so their combined cost is £12. That's not right: the question said that the combined cost is *£11.*

When you realised that £1 was the wrong answer, did you pause before reading on and try to work out the actual price? Or did you react as many teenagers do: 'it's not possible' or 'it's a trick' or even 'this reminds me of why I don't like maths'.*

These negative reactions are very common, and they give an important clue about the difficulties that many teenagers have with maths.

There's some psychology here. The Nobel Prize winner Daniel Kahneman has found strong evidence that our minds operate on two systems, one that is fast and intuitive and a second that is slower and deliberate. Fast thinking is about intuition, and this type of thinking tends to dominate, even though we don't always recognise that this is the case. Slow thinking involves being analytical, and requires increased effort and concentration.

The Bow and Arrow puzzle is an example of where our fast intuitive thinking takes over, but turns out to be wrong, and so we have to switch over to slow mode to figure out the solution. Most

* The correct answer, in case you didn't find it, is that the bow costs £10.50 and the arrow costs 50 pence, it's discussed further on page 108.

of us spend much more time thinking fast (intuitively) than we realise, and find thinking slow (analytically) a challenge. This is even more the case for teenagers, and the constant battle between fast and slow that maths forces on them is one reason why they find it so hard.

So what can parents do about it?

First, being aware of the way that they're thinking is a great help to working with them and lessening 'that sinking feeling'. But it's also important to realise that there is a positive side to slow thinking. Many people report that when they get engaged in activities like maths or writing or painting they cease to be aware of the passage of time – what's sometimes referred to as a state of 'flow'. And research suggests flow states can lead to a tremendous feeling of satisfaction.

Half the battle here is getting started – because, deep down, we know doing hard thinking – like maths homework – isn't going to feel, initially, that pleasant, so we put it off. But once over that hurdle, flow can kick in.

So here is our first suggestion. Rather than banishing your teenager to the bedroom until their maths homework is finished (never likely to work at the best of times), it may be better just to get them to agree to work on it for ten minutes. That ten minutes 'cranks them up' and, perversely, is more likely to put them into the frame of mind to continue.

The Bow and Arrow question is an important reminder that, for most people, maths relies more on effort than on innate ability.

It's normal for teenagers to feel some resistance to maths and it's better for parents to acknowledge that and help their teenager work with, and through, the resistance than to try to use delaying techniques and push it to one side.

And don't assume that all teenagers dislike the difficulty involved in some parts of maths. As one fourteen-year-old girl (of average ability) said to us:

I like maths being hard, challenge is good.

HOW PARENTS CAN HELP

Mum – I think you need to date a maths teacher.*

Many parents find that their own direct efforts to teach maths don't work, for one important reason: most teenagers regard their teacher as the authority on everything to do with maths (regardless of whether they like the teacher or not). The teacher is 'right' and anything a parent does that appears to contradict the teacher, or which uses a different method, is likely to be ignored.

This means that what you do at home has to complement rather than compete with what is going on at school. And you have to be subtle, too. Not only does it often help to use the word maths as little as possible, in many situations it can be more effective if you remain at arm's length (unless you are specifically asked to help). What you can do as a parent is to set up the opportunities for maths engagement, without necessarily having to be there when they happen.

We've talked to parents of teenage children of all abilities and backgrounds, and in doing so there are three broad themes that

* Advice from a thirteen-year-old girl on how her mother can help with her maths.

come up again and again. We've therefore split this chapter into three sections, though it's worth saying that all three themes are related:

1) What's the point?
2) Motivating your teenager
3) Getting a good exam grade.

What's the *Point*?

'What's the point?' and its equivalent, 'When am I ever going to need this?' are the killer questions to which every parent would love to have an answer (and many teachers too).

Some parts of maths are more prone to these questions than others. Maths up to GCSE can be crudely divided into three types:

- Practical everyday maths, such as arithmetic, percentages and statistics. This maths has such a clear application to everyday life that its value will rarely be questioned even by the stroppiest teenager.
- Maths that is only relevant in certain professions after leaving school. Algebra and some parts of trigonometry are essential for those who want a career in technology and science. Engineers and computer scientists will use this maths regularly. But there are many other professions, even scientific ones like medicine, where it's unlikely this knowledge will ever be called upon. If your ambition is to be a chef, it's reasonable to question why you have to learn about simultaneous equations.
- Finally there is abstract and pure maths. Circle theorems, cosine rules, factorisation, solving a quadratic equation,

and the concept of mathematical proof . . . the chances of these ever cropping up in anything except a maths question are exceptionally small. The fact is, most teenagers *aren't* ever going to need to use these things – as their parents can tell them, having last encountered them when they themselves were at school. For these topics, the 'point' is much more subtle, and if you are faced with having to justify it you have a much harder task.

So as a parent, how do you explain the point of these higher levels of maths?

Let's start with the hardest part to justify, the abstract side of maths that has no obvious applications to real life.

Three standard arguments for doing abstract maths are these:

- This maths teaches you how to reason logically, to problem-solve and to be rigorous. This is training in how to think, and no other subject does this quite so effectively. In later life, you'll be grateful that you learned how to think in this way.
- Understanding this more abstract maths strengthens your understanding of the more routine everyday maths. You may not use the maths of (say) circle theorems directly, but it can build your confidence in the maths that you do use.
- Sometimes this 'useless' maths turns out to be useful after all. For example, some of the most abstract research into prime numbers has turned out to be crucial in finding ways to encrypt information on the Internet, something that the world economy now depends upon.

All of these claims are true. Unfortunately, they rarely convince a teenager.

More convincing, though rather defeatist, is the argument

where the parent just comes clean: *Yes, it is not immediately useful and you may never need it again after your exams, but like it or not you have to get a maths GCSE to get a decent job, so just grit your teeth and get on with it.*

But this all sounds very defensive. Let's step back and look at *why* teenagers ask 'What's the point?' in the first place. In our experience, teenagers are often not that bothered about whether the stuff they are learning is going to be useful when they are an adult. Indeed many parents complain that their teenagers don't think about the future enough, because they are so focused on simply enjoying the present.

We reckon that the question, 'What's the point?' is almost always an afterthought. It is usually the consequence of the teenager being either stuck or bored. It is when the maths they are faced with is either too difficult or is unstimulating that their minds turn to wondering why they should be doing it in the first place.

We met a surprising number of teenagers whose favourite part of maths turned out to be algebra. When asked what the point of algebra was, most didn't know. All that mattered was they could do it, and that they liked the feeling of, 'Ah, now I get it!' and 'Yes, I got the right answer!'

So the secret is really in finding a way to help your teenager to succeed at the maths they are doing – to be challenged, but not too much.

As a parent, there is a simple but crucial word that can make all the difference here. The word is: *YET*.

When maths becomes challenging, instead of the reaction 'I can't do this,' you should encourage the idea of 'I can't do this *yet*.' This simple phrase can help a lot with engagement in maths, since it conveys the idea that maths is something that involves progress and personal growth, rather something you either can or can't do.

The teenagers who begged for algebra lessons

A nice story of teenagers being driven to learn maths comes from Homer Hickham's autobiographical story *The Rocket Boys*. A group of teenagers in 1950s America got their kicks from firing rockets. Disappointed that they weren't achieving enough altitude, they decided to figure out how to make their rocket go higher. One of them found a book about rocket engineering:

> *I passed the rocket book around, inviting all the boys to inspect the pages of equations. 'To get all that we need to know from this book,' I said, 'we're going to have to learn calculus.'*
>
> *'And differential equations,' added Quentin.*
>
> *'Are you two crazy?' Roy Lee demanded. 'We can hardly do the homework they give us now!'*
>
> *'Nevertheless,' Quentin said. 'It must be done.'*
>
> *Roy Lee sighed. 'Here we are, a bunch of West Virginia hillbillies wanting to be Albert Einsteins.'*

When maths helps you to pursue one of your interests, it suddenly has a point. Context can be everything. A teenager who enjoys card games will find probability fascinating. If they are into cooking, ratios will be extremely relevant, while many geometry ideas crop up in design and other artistic hobbies. The power of computers and the web has created many opportunities here. Many teenagers are designing websites or learning to write apps (sometimes to sell) and mathematical thinking is involved in all computing. Some teenagers have got sufficiently enthused by a mathematical puzzle or trick that they've even decided to create a video for YouTube to explain it to others. As a parent, it's hard to influence your teenager's interests, but you can look out for where maths fits in.

Motivating your Teenager

There are many reasons why teenagers may lack motivation to do maths. Some, such as not being able to see the point of what they are doing, we discussed in the previous section.

If your teenager is reluctant to engage with maths, here are some ways that might get them more motivated.

Role models.

Role models can be a powerful influence. Ten years ago, it would have been a struggle to name anyone famous or 'cool' that likes maths. Today the situation is very different. Some have called it 'The Brian Cox' effect, though in fact when the media calls on somebody to talk about numbers there are now dozens of media-savvy people who can do so in an accessible way. Maths is even getting linked to comedy. Dara O'Briain and Dave Gorman are two top comedians who are proud of their maths background. Maths may not be universally cool yet, but it is considerably more so than it was, and teenagers are beginning to realise this.

Making (and losing) money.

Teenagers are more interested in money today than they have ever been. Maths that will help you to make money is a strong motivator, and plenty of studies have found that somebody with a maths qualification will earn more in their career than somebody without that qualification. Just as important as making money is knowing how not to lose it. The 'Monthly Allowance Challenge' below is a great example to try out on your teenager.

Knowing which careers involve maths.

Sometimes the problem is ignorance of just how many careers depend on maths. There was a time when the careers adviser at school would say that with a maths qualification you had the choice of being a maths teacher or an accountant. These days, it's more a question of finding those few professions where you *don't* need maths. There is an excellent, long-standing website that will help: www.mathscareers.org. Also, there is an online maths magazine for teenagers, *Plus*, which regularly features interviews with people who do maths for a living.

The monthly allowance challenge

If your teenager takes an interest in money (and most do, or at least the idea of making lots of it), try this. Tell them you are thinking about making a contribution next month to their leaving-school fund (or something else that you're expecting to pay a lot of money for in the future). Say that you're wondering what contribution to make, and there are three options:

- Make payments every day for a month. You will pay them a penny on the first day of the month, 2p on the second day, and keep on doubling it each day until the end of the month.
- A simple flat sum of £200.
- Payments of £45 for four years, increased each year to allow for inflation.

Which deal would they prefer?

Most teenagers will take either the £200 or the 'index-linked' payments of £45, though they might at least be tempted to know what the first offer of the doubling penny is worth. If they are, tell them to work it out for themselves. If they bother to do so they will get a huge surprise. If (as is more likely with most teenagers) they don't bother to work it out, agree to the lump sum payment, and then as an aside point out what the first offer was worth.

The first seven days suggest it will be a pittance: 1p, 2p, 4p, 8p, 16p, 32p, 64p. The second week is more interesting: £1.28, £2.56, £5 (rounded down), £10, £20, £40, £80. That £200 isn't looking so generous now.

Week three gets exciting: £160, £320, £640, £1280, £2500 (rounded down), £5k, £10k. And week four – well, this would mean selling your home, cashing all your savings and taking out a massive loan: £20k, £40k, £80k, £160k, £320k, £640k, and on day twenty-eight (the last day of the month if it's February) £1.28 million. We've yet to find a teenager who wasn't impressed by this, and it's an excellent lesson in the importance of knowing how numbers work.

Games

It's no surprise that maths educators have spotted that games are one of most effective ways of reaching teenagers, given the number of hours that many of them are planted in front of a screen playing at home. Not all games involve useful maths, of course. You could argue that when playing computerised football there is some maths involved in working out which angle to pass the ball to the winger, but, if so, then it is all intuitive and has little to do with maths that might be needed in any other situation. On the other hand, there are a number of computer games and activities that teenagers enjoy that require an ability to answer explicit maths questions. These include, for example, *BBC Bitesize*, *Mangahigh* and *Professor Layton*, popular games that require them to solve maths questions, usually in the form of puzzles that have to be tackled before you can progress to the next stage. You can ask other parents or your teenager's teacher if they have any more suggestions.

Use real, everyday-life problem-solving

Nothing beats learning maths when you – or in particular when a teenager – actually need it. If they are planning a party, or going to a festival, or a trip somewhere with friends, they will need to budget for it and you can help to lead their thinking. The tactic of 'thinking aloud' can sometimes work, even though it's a one-way conversation. As an example, suppose you'll be making a contribution for their party or an outing to the cinema. You can do your budgeting mentally out loud: 'Let's say there will be six of you . . . the price of tickets is £9 . . . let's call it £10, so that's £60 . . . and you'll probably be wanting snacks, let's say £3 each . . . plus your return bus fares, £2 each . . . so that's £90 . . . I'll pay half of that, £45, so that leaves you £45 to pay between the six of you, £7.50 each.' This is almost inevitably a conversation with yourself, though you might pause before answering all of the questions, just in case your teenager feels induced to chip in on your behalf. Even if they don't, despite the displays of indifference you might spot, there's an important subliminal message being conveyed – that people do actually use maths to work things out.

Expose them to *interesting* maths

Many teenagers disengage from maths not because they don't have the skills, but because they find it incredibly dull. That may be a reflection of the particular teacher that they have, or of the curriculum, which in the wrong hands can resemble a set of drab, repetitive exercises solving problems that bear little relation to the teenager's interests. We've seen many examples of teenagers who find maths dull and yet have been inspired by a) a TV programme, b) a talk by a speaker (at their school or at a science festival) or c) very occasionally, a book.

Maths that seems to particularly grab the attention of teenagers generally involves:

- Humour (a maths problem involving characters with funny names will be far more appealing than the same problem set in a dull everyday context).
- Connection to a subject they are particularly interested in (such as money, a particular sport or an issue like global warming).
- A sense of 'Wow' (in the form of a magic trick, for example).

The quickest way to find material like this is to have a scan on YouTube. There are dozens of people who have found ways to make maths intriguing, accessible, funny and even cool, producing material that would never make it on to mainstream TV. Search YouTube for videos by, for example, Vi Hart or Numberphile, and see where they take you.

Give them a numerate task to do

If you are at all familiar with spreadsheets then get your teenager to help you with them. Or better still, get them to set up a spreadsheet for you. The sort of thinking behind solving the problems posed by getting a spreadsheet to work is exactly the sort of thinking behind understanding algebra. If your teenager finds a part-time job involving numbers, that can be even better. Many maths skills are learned on the job, sometimes just through exposure, and sometimes motivated by the shame of not being able to work something out when the 'boss' asks for it. We heard of one teenager who was working in a shop at weekends, who returned home to beg his parents for a lesson on how to work out a percentage discount because he'd been given a roasting by the manager for not having learned it at school. (He'd presumably been exposed to percentage discounts at school as it's a standard part of the curriculum, but it clearly hadn't stuck.)

Rewards and praise

Praise is a strong motivator. Even a sullen teenager who responds in nothing more than grunts will be inwardly registering when they've been told they have done well or badly. Praise is good for self-esteem, but it can also backfire in some surprising ways. Studies in recent years have shown that children as young as four begin to tell the difference between praise that is deserved and praise that isn't. Indeed, in many cases children begin to observe that those with lower ability tend to get more praise as a form of encouragement to keep going, and they therefore associate it with 'you aren't doing very well' – the complete reverse of what it should mean. As a parent, therefore, it's important to ration praise and use it appropriately, rather than dishing it out at every opportunity.

Then there is the question of what a child should be praised *for*. Most parents are naturally tempted to praise their child for being clever. Cleverness is a quality we all like to have and it's nice for it to be recognised. Unfortunately, all the evidence suggests that praising a child for being clever is one of the worst things you can do for their motivation. Extensive research by psychologist Carol Dweck (and confirmed by others) has shown that children whose intelligence is praised (through repeated use of phrases like 'clever girl' or 'clever boy') become anxious that at some point they are suddenly going to lose this quality. They become afraid of failing, and as a result they often stop trying. Some even begin to find ways of faking their 'intelligence' by cheating. In one famous study, able children praised for being intelligent actually performed worse in tests than children who had achieved lower scores than them and had not been praised.

The message is clear: while intelligence is something that a parent can feel quietly proud of, what should be praised is effort, particularly when it comes to maths because maths is a subject where effort and practice pays off. Don't confuse this with some sort of dumbing down: we're not saying that intelligence or high performance is bad. We're simply saying that in order to get high performance in maths, hard work is even more important than natural flair for the subject.

Many parents, frustrated by the apparent lack of commitment of their teenagers to maths, resort to financial rewards. *If you get an A, we'll get you that bike that you wanted*. But the pressure of being rewarded for high grades can be almost as detrimental as praising intelligence. Some schools give pupils grades for performance and for effort, and it's the latter that parents should be looking to reward – because in the end, that is what will improve the former. And ironically, psychology studies* found that although shifting attention onto getting an external reward rather than the satisfaction of a job well done might have a short-term impact, it actually *discourages* teenagers from wanting to go on and do more maths.

Helping Them to Get a Good Exam Grade

In an ideal world, the main concern of every parent would be that their child receives a maths education that leaves them inspired, curious and with a practical grasp for the subject and its applications.

But we'd be doing a disservice to the majority of parents if we ignored what is, sadly but inevitably, an even greater priority than enjoying the subject. Namely, exam results. As we've mentioned already, the debate about whether some maths is useful or not is largely irrelevant. A decent grade in maths is now deemed to be a requirement for a vast range of courses and careers. To get into nursing, graphic design, retail, even (in one case we heard of) a job in a call centre, employers are almost certain to want to see maths in an applicant's qualifications. It's not that they require you to be able to work out square roots or the area of a triangle (there are hardly any jobs that require direct use of either of those maths skills), but a maths qualification has become one of the essential passports to entry into most aspects of adult life. Faced with a huge number of

* For good examples read *Punished by Rewards* by Alfie Kohn.

applicants, interviewers use the lack of a good maths qualification as an effective filter for excluding many candidates who might otherwise be up for the role. So it's understandable that almost every parent regards getting a good maths grade as a priority for their child.

Parents are not alone. The government also wants to see as many teenagers getting good passes in maths as possible. Indeed, what governments like is to see the proportion of good grades increasing every year, since this creates a convincing impression that their education policy is working. And because government cares so much about exam success and sets targets, schools care as well. This is why many teachers will tell you (especially off the record) that the main objective of maths lessons is to train pupils in how to achieve higher grades. Whether they get a good maths education in the process, acquiring skills that will be useful in later life, is often less of a concern.

This would be fine if it meant that our nation is getting better at maths, but there is plenty of evidence to suggest that teenagers aren't getting better at maths, they are simply getting better at completing the exam papers.

How can you get your teenager through the exams if they appear to be struggling with maths?

There are revision books, tutors and online games that might get them over the line. Teachers are often adept at predicting which questions will come up in the next exam, and there are various strategies that you might come across that have been successful in getting pupils a pass grade.

One common technique is selective 'cramming'. When two levels of exam are available (Foundation and Higher) schools will enter borderline students for the Higher exam and coach them intensely in a narrow range of topics. A score of 30 per cent in a

Higher exam might get the same grade as a score of 70 per cent in a Foundation exam.

If getting a grade is all that matters, then these strategies can work, but there is a serious risk that your child will emerge with a certificate but minimal understanding or skill. They are also likely to have had a pretty lousy experience of maths, and will therefore go on to perpetuate the negative views that many adults have about maths in school.

The other problem with cramming for the exam is that it is not good preparation for those who are going to take more exams later on. Students often get a nasty shock when they start maths in the sixth form and discover maths that requires a much deeper understanding than they needed to pass GCSE.

Cramming and memorising do have their place, but the path to long-term success in maths exams involves, more than anything else, perseverance and practice. Doing well at maths means doing lots of it. And to do lots of maths, a teenager needs to find it engaging – they need to be able to understand it, and to have some success at it. And that is where the rest of this book comes in.

HOW HAS MATHS CHANGED?

The Syllabus

If you listen to the media, you might be under the impression that maths has changed out of all recognition from previous generations. And yes, the syllabus has changed, though perhaps not as much as you might guess.

Certain topics that became fashionable in the 1970s have largely disappeared. For example, matrices and set theory have gone, and most pupils will now never encounter these unless they go on to read a mathematical subject at university. The biggest growth has been in probability and statistics. Many older parents never touched probability at all in school. Today it is rightly recognised as being an essential tool in understanding the world around us and the decisions and risks that we take.

There is now a greater emphasis on what is called *functional maths*. This is an attempt to extract from maths those parts that are connected to real-life problem-solving. However, it has proved very difficult to draw the line between which parts of maths count as 'functional' and which parts don't.

The syllabus is a political hot potato and constantly under review. Over thirty years, the examination system has experimented with introducing coursework (scrapped because it was found parents were often doing the work) and 'modular' exams (phased out because topics were crammed and then forgotten). Accusations have flown around for years of maths being dumbed down, lacking rigour, not being relevant enough, being too prescriptive, and so on.

The truth is, however, that good teaching can foster good maths learning regardless of the curriculum, and poor teaching will leave learners struggling whatever is supposed to be taught.

The Classroom and Technology

Calculators and computers have also had a significant impact on the content of maths lessons, particularly by reducing the amount of time that is spent in class doing calculations. It's hard to justify spending hours doing long-winded multiplication and division when these can be done almost instantly with a calculator.

But walk into a typical classroom and the biggest change that you will notice is the absence of blackboards. Writing with chalk is a dying art.* In its place has come the interactive whiteboard, which at its most basic is often used like a blackboard, though one where you can save and print off all of the pages that you write up, rather than having to rub them off. More important, however, is that these whiteboards are used as computer screens, allowing teachers to rapidly plot graphs or highlight geometrical patterns that in the past would have required minutes of careful scribing. This means that many teenagers now have a much better visualisation of maths than previous generations. They can

* Chalk and blackboards might be disappearing in the UK, but most schools in continental Europe still use them. In terms of technology in the classroom, Britain is surprisingly advanced.

rapidly plot the results of an equation, and then see what effect changing different numbers in an equation has on the graph. In the right hands this means they have far more opportunity to explore maths ideas than their parents ever had.

You may find that some of technology being used in school is seeping into your home, too. Maybe your teenager already submits their homework via a website; maybe they use online software like GeoGebra to learn about geometry.

There can be a downside to this maths technology. In the days of blackboards, while chalking up patterns on the board the teacher could talk through what was going on. Now that everything can be produced instantaneously, some learners don't have enough time to process what is going on. Thinking time is very important in maths.

New Terminology

No doubt you will encounter some unfamiliar words in maths, some of them long-established words that may not have featured in your syllabus (such as *surds* and *tree diagrams*) and others that describe 'modern' techniques (such as *chunking* for division). All the important terminology is covered either in Part Three of this book or in the Glossary at the end.

If you do only five things:

- If your teenager asks for help, resist the temptation to broaden it into a lesson on the bigger picture or to go back to first principles (unless they say they want that, which is possible though rare). Instead, find ways to introduce the big picture in situations away from homework.
- Remember that you are complementing rather than competing with what is going on at school. Recognise that their teacher might be teaching it one way, and that it may not help to show them another way.
- Prepare before giving advice. The parent who launches into helping with homework then gets stuck and becomes diverted onto working it out him/herself can drive teenagers mad.
- Remember that maths doesn't have to have a point! If you're engaged, or laughing, or succeeding, then anything from solving quadratic equations to memorising the digits of pi to ten decimal places can feel like it's worth doing.
- Don't let your teenager hear you boast that you are hopeless at maths. This is a British disease – in no other country is it so acceptable or so common to hear adults make this claim. To a teenager it sends out the signal that maths doesn't really matter, and it's a phrase that drives teachers up the wall. Part of the problem is that it isn't even true: many parents who say they are hopeless at maths are extremely competent at it. 'I'm hopeless at maths' is usually a defence mechanism for anyone who isn't as quick at mental arithmetic as the experts on *Countdown*.

PART TWO

THE EVERYDAY SKILLS

BUY 3 TYRES FOR THE PRICE OF 4 AND GET THE 4TH ONE FREE *

Let's forget about the school curriculum for a moment. What do you want your teenager to emerge from school with (apart from a good grade, of course)? It is important to have this big picture, because you can keep coming back to it in those moments where you are bogged down in the detail and you and your teenager are wondering what the point is.

Of course it is essential that your teenager leaves school with skills that will make them employable. But they also need to be able to function socially in a world that increasingly requires them to handle numbers – and in a world that is forever looking for new ways to rip them off (as the advertisement at the top of this page illustrates).

We reckon that there are three broad maths skills that are essential for anyone who wants to thrive after leaving school:

1) The ability and confidence to tackle unfamiliar problems.
2) A good feel for numbers, including the ability to estimate without needing a calculator.
3) A solid understanding of ratios and proportions (including percentages).

* Advertisment seen outside a garage.

These might seem a strange combination, ranging from the very general to the very specific, but actually the three are all connected. Each one is a fundamental part of linking the maths your teenager does at school with the situations they are likely to encounter in adult life.

PROBLEM-SOLVING

Teacher: Here are some timetables for the Manchester trams. Can anyone tell me how long it takes to get from Altrincham to St Peter's Square?

Pupil: We've only done trains, we aint done trams yet, miss.*

Employers and universities consistently complain that their recruits lack the ability to solve problems and work independently with unfamiliar information. It seems that far too many teenagers emerge from school unable to transfer the techniques they learned in maths lessons to anything that isn't a question of the type they have been taught.

This is why doing maths in the form of problem-solving is so important. Life doesn't come along and say, 'I am a maths problem involving simultaneous equations with a nice whole

* Overheard in a classroom.

number answer.' Sometimes it's not even obvious that it's a maths problem at all.

Most school maths tends to reinforce the impression that maths is a matter of learning a series of disconnected techniques. Yet the truth is that at its heart maths is a problem-solving activity; indeed the whole history of how the subject developed is connected to solving real, often life or death problems. The Egyptians needed to be good at maths because they wanted to find more efficient ways of planting crops and building pyramids. Sailors and explorers needed geometry in order to navigate in unfamiliar lands. And many now believe it was mathematicians who won the Second World War, thanks to the secret code-breaking work at Bletchley Park.

If teenagers were more exposed to mathematics as problem-solving rather than as the memorisation of a series of separate, individual techniques, then they might also be more interested in the subject – *and they might be more employable, too*.

Interesting mathematical problems resembling puzzles sometimes crop up in real, everyday situations. Take this example, which genuinely cropped up one wet Sunday afternoon.

> Twelve-year-old Amy had been given an old 300-piece jigsaw, now kept in a tub instead of its original box. She tipped the pieces out on the floor. As always when doing jigsaws, she started by looking for all of the edge pieces of the puzzle. As she hunted around the box, she noticed there didn't seem to be enough edge pieces for a jigsaw this size, and she wondered if maybe some had been lost. She counted up the edge pieces, but realised she didn't have anything to check it against. How many edge pieces should she have expected to find?

There is nothing on a jigsaw box to tell you how many edge pieces there are, and Amy was stuck. Being stuck is a sign that this is a genuine 'problem'. You might like to think a bit further about how you would begin to tackle it before reading on.

A moment of insight comes when you realise that the total number of pieces in a jigsaw can be worked out by multiplying together the pieces along the top by the pieces on the side. Since this is a 300-piece jigsaw, there could in theory be several combinations that would work for pieces along the top and side, for example: 50 × 6; 30 × 10; 25 × 12; 20 × 15. (These are known as the factors of 300, more on page 97.) On the other hand, there couldn't be, say, 22 pieces along the top of the jigsaw, because 22 doesn't divide into 300.

Now comes the practical element. If you think of a normal jigsaw, its shape is usually fairly square. Long thin rectangular jigsaws are very unusual. Look at the possible edge combinations. Could Amy's jigsaw really be 30 pieces along the top and 10 down the side? When did you last see a jigsaw with those dimensions? The only combination of sides that multiply to make 300 and that resembles a typical jigsaw is 20 × 15. How many edge pieces are there on a 20 × 15 jigsaw? 20 along the top and bottom, plus 15 along each side, adds to 70. But wait, the corner pieces are on two sides, and they shouldn't be double counted. So the actual number of pieces will be 70 – 4 = 66.

Solving this problem involves a couple of mathematical discoveries, but it also brings out another point, which is that real-life problems require a dollop of common sense. Yes, it's possible that the jigsaw was 30 pieces along and 10 high, but it would be a very unusual-looking jigsaw. If you had to take a punt on the right answer, you'd say 66 pieces. (And in this case you'd be right, since this was a true story and Amy did indeed find 66 edge pieces.)

Much of the skill of problem-solving comes with age and experience, but there are certain principles that as a parent you can help to nurture. Here are some of the most important:

1) **Make sure you understand what the problem is in the first place.** It doesn't matter how hard you work on a problem, it counts for nothing if you aren't solving the right problem. Teenagers often learn this the hard way, in exams, where they fail to read the question. Part of the problem is the teenage tendency to want to get things over and done with as soon as possible. There's not much a parent can do to prevent this. However, the second reason why teenagers go off in the wrong direction is that they are embarrassed to check they have actually understood. If your teenager wants some help with some problem, one of the most important things you can help with is ensuring that they have understood what the problem is. Often the biggest challenge is turning a wordy problem into mathematical notation – it takes a lot of practice. A good discipline is to read each piece of information in the problem and try to write it down as a mathematical expression before moving on to the next piece of information.
2) **Experiment by trying an answer that is probably wrong, and see what happens.** Faced with a difficult problem, most people have the immediate reaction, 'I can't do this.' This is even more true for most teenagers. Yet there is plenty of evidence that the best problem-solving comes from people who are prepared to get their hands dirty and try things out, and who regard 'failure' as a chance to discover more about the problem. It takes confidence to overcome this huge barrier, not least because 'having a

go' can mean admitting you don't understand. As a parent, having a go first and making a mistake (and being OK with that) can give your teenager confidence that they can try the same thing.

3) **Start with a simple example where you can see what is going on, and build up from there.** Even the most able people can find difficult problems . . . difficult. But there is one skill that good problem-solvers use all the time, which is taking the initial problem and simplifying it. Lead by example. Say: 'That sounds too hard, let's imagine a simpler version . . .' This could involve changing the numbers involved, choosing simpler ones to work with (for example, if trying to find the number of rectangles on a chessboard, start with a 4×4 board rather than an 8×8).
4) **Try drawing a diagram or a quick sketch.** Often the wording of a problem can be confusing. Turning it into a visual representation can make clear what is being asked. Or draw up a table, or try to express the question differently. All these get the problem 'off the page' and help your teenager take ownership of it. Incidentally, many teenagers seem to think that drawing a diagram is almost an admission of failure, and that somehow the 'right' thing is to be able to sort out a problem in your head. Help them to get over this embarrassment by showing them that scribbling diagrams to solve problems is what good problem-solvers do. Hey, Leonardo da Vinci scribbled diagrams *everywhere* in his notebooks.
5) **Check your solution to see that it makes sense.** If ever there was an almost universal flaw in teenage problem-solving it is this one – especially when it comes to maths problems. Such is the relief at coming up with an answer

that most teenagers can't wait to move on to the next thing. There's also an element of denial. They don't want to check the answer in case they find something wrong, which means they have to do more work. In fact, it might not be a bad thing to let them move on to something else, as long as they are encouraged to check back on the original problem later. Often allowing a gap of time between completing a problem and checking it can reveal mistakes.

Two other tips are:

6) Be systematic, and only change one thing at a time.
7) When seriously stuck, take a break.

Of course you can read tips like these until you are blue in the face, but you only really get to be good at problem-solving by trying it out for real.

We have put together a collection of problems and puzzles (page 59). Between them they should bring out all of the problem-solving tactics that we discussed above. None of them requires any advanced maths, but that doesn't mean that they are easy. In fact, some of these have been known to take people several hours to solve. If you *do* try one of these problems with your teenager, resist the temptation to look up the answer in the back. The feeling of being 'stuck' is an important part of problem-solving, and one that your teenager needs to experience and to become comfortable with if they are to be confident problem-solvers later on.

PROBLEM-SOLVING

If you do only three things:

- Some 'real life' problems can be easily solved using maths language. The hardest part for your teenager can be working out how to express word problems in the language of maths. Puzzle-solving makes great practice as does encouraging your teenager to put things into their own words.
- Solving a problem means being prepared to get stuck and go up blind alleys. It's important to reassure your teenager that mistakes are inevitable and OK (lead by example!).
- Find problems that your teenager cares about. If a question makes them think 'who cares?', it's unlikely they'll put much effort into tackling it.

ESTIMATION AND A FEEL FOR NUMBERS

We're not talking about quantum physics here, are we? We're talking 'this rose cost 40p, so if I take ten roses that's £4.' *

Most people think of maths as a very precise subject with 'right' and 'wrong' answers, so it is a touch ironic that one of the most important mathematical skills is the ability to come up with answers that are only *vaguely* right.

The ability to estimate, to do rough workings on the back of an envelope and to have a feel for whether an answer looks roughly right or not is an essential life skill, whether you're working out the best buy for a mobile phone or trying to make sense of the statistics that are in the news. In fact, we would argue that in adult life, the ability to come up with an approximate answer is likely to be useful just as often as the ability to produce the exact answer. What's £28.40 × 1213? If you're doing the accounts then a

* Lord Sugar, exasperated at the innumeracy of *Young Apprentice* candidates in the show's second series.

calculator or spreadsheet can work out the right answer for you. But it's usually going to be more important that you know the answer is (very roughly) 30 × 1200, which equals 36 000. Otherwise, how will you know if your finger has accidentally pressed a calculator button twice and given you a spurious answer? (We talk more about the problems with calculators on page 253.)

We spoke to a number of employers, and the message came back time and again that recruits often lack a 'number sense' even if they have a GCSE in maths.

So, if all you really need to know after leaving school is times tables and a few back-of-the-envelope methods, maths begins to sound very simple. However, making the maths simple is actually a skill that requires a lot of experience. An adult with modest maths ability is generally better at estimating answers to real-life problems than a teenager who is really good at maths. You need to know when an answer really is 'close enough', when a certain factor can be ignored, and so on.

If you are going to become good at estimating, there are three number skills that over a lifetime will handsomely repay the effort invested in learning them:

- Know your tables. If you need a calculator to work out 4 × 9 then you'll struggle to deal with many impromptu situations life throws at you. (Not many people whip out their calculator in the supermarket.)
- Know how to round numbers. This isn't a skill that is necessarily used on its own, but it is central to one that follows.
- Know how to multiply numbers ending in zeroes. Most important is multiplying by tens and hundreds, since these are the most common, but the ability to multiply by larger

or smaller numbers is also important. Scientists tend to use what is called 'standard form', where for example 8000 is 8×10^3 . More on page 210.

The mental calculation test

We asked a cross section of twelve- to fifteen-year-olds from a range of different schools the following question. They were not allowed to use a calculator:

You buy 80 chocolate bars for 70 pence each.
How much do you spend in total?

Most of them recognised that this question involves multiplying 7×8, and the majority knew that this is 56. The problem came in turning this into the correct number of pounds. The most common answer among young teenagers was £5.60, their reasoning being that this sounds 'about how much you'd expect to spend on a lot of chocolate'. Only about a third of twelve- to thirteen-year-olds got the correct answer of £56, though this proportion increased significantly for older age groups, which suggests that arithmetic ability combined with a common-sense understanding of the world tends to increase with age. Even so, answers of £560 were not uncommon – and a significant minority came up with answers such as £5.90, or £57. Our conclusion, though, is that while teenagers' knowledge of tables seems reasonable, their ability to handle the equally important zeroes and decimals can be very shaky indeed. Rounding the figures here to more convenient amounts provides an easy way to check the reasonableness of answers. If the chocolate bars were £1 each then 80 of them would cost £80. At 50p each the total would be £40, so the answer must be between these two figures, £40 and £80.

Zequals – The Estimation Game

Zequals is the ultimate game of estimation. If your teenager leaves school able to play it, they will have an extremely powerful skill in their toolkit, since they will be able to estimate the answer to any calculation in their head. This is done by making each calculation as easy as possible by brutally rounding every number involved at every stage.

To turn a 'hard' number into an easy one, use *Zequals*. We've borrowed a symbol usually used in astrology (the Aquarius symbol) and called it the Zequals symbol.

♒

It is like an equals symbol but with zigzag lines to show it's only *roughly* equal and all those Zs hint at the final answer being likely to end in zeroes. (You can find the symbol if you use the Wingdings font on your computer – it's what comes out when you type an 'h'.)

The idea of Zequals is to find the nearest round number that is a single digit followed by zeroes. Technically this is known as a number rounded to *one significant figure* (see page 208). It sounds complicated, but with practice it becomes second nature.

So: 23 ♒ 20
65 ♒ 70 (when rounding a 5, always round it *up* to the nearest 10)
118 ♒ 100
563.5 ♒ 600

The numbers 1 to 9 stay the same when you use Zequals, so for example 1 ♒ 1 and 8 ♒ 8. Decimals are also rounded to one significant figure, so 0.43 ♒ 0.4 and 0.089 ♒ 0.09.

Now for some calculations.

Using Zequals, $93 - 17$ is simplified to $90 - 20 = 70$

$7 \times 9 = 63$, but in Zequals, 63 is simplified to 60,
so $7 \times 9 \approx 60$

The rule is that *whenever* you come across a number that has more than one non-zero digit, you Zequal it before going any further.

Try a harder calculation: 77×61. When we give it the Zequals treatment it becomes: 80×60.

Multiply those numbers together to get: 4800.

And even this answer can be Zequalled, since we just want a single digit followed by zeroes. So $4800 \approx 5000$.

In other words, in the game of Zequals, $77 \times 61 \approx 5000$.

Despite the fact that we have done some extreme simplifications here, 5000 is not that far from the precise answer, 4697. It shows that rough estimation, when done properly, is often remarkably accurate. In many situations in life, Zequals is as much maths as you need. The Zequals game can be played when doing any calculation, particularly when tackling Fermi questions (see opposite). How many spectators are in that football crowd? How much money does this café make every week? When will the world run out of oil?

We should add a caution that Zequals is not a technique that is formally recognised by examiners (not yet, anyway). Exam questions make it clear what level of rounding will be needed in an answer. However, Zequals can be of use even in exams, and certainly in the classroom, as a method for checking that an answer is in the right ballpark. If you come up with the answer 89.6 and Zequals says the answer is about 1000, you can be fairly sure that you put a decimal point in the wrong place in your calculations.

Incidentally, you're never too young to start to learn Zequals. We've known children aged eight confidently able to use Zequals to multiply 13×48 (answer $10 \times 50 = 500$). Even six-year-olds can do the first step, of turning any two-digit number into its nearest multiple of ten. 41 ≈ 40 and 88 ≈ 90. The earlier they start using it, the more comfortable children will be in using it to make estimates.

TEST YOURSELF

Have a go at these calculations using Zequals:

a) $19 + 63$
b) 38×12
c) 1.84×97
d) 436×68

Fermi Questions

The master of 'Back of the envelope' estimation was a physicist called Enrico Fermi. Fermi was involved in the development of the first atomic bomb, and developed a reputation for producing accurate estimates using simple calculations. One famous example was the testing of the first bomb in the Nevada desert. Fermi and his colleagues were watching the explosion from a position several miles away. As the bomb exploded, it is said that Fermi held up his arm and let drop a handful of paper confetti. The gust of wind from the explosion blew the confetti several feet away before it landed, and based on how far they travelled, Fermi estimated the strength of the bomb. It turned out that his estimate was about half the correct answer. Normally, if your answer is this far out it gets zero marks, but in this case being 'only' out by a factor of two was a remarkable mathematical achievement.

Fermi questions are a good way of practising estimation, and if your teenager needs any motivation, it's worth pointing out that many employers use Fermi questions in interviews as a way of seeing whether you can think on your feet. Want to work at Google? They'll probably ask you a Fermi question at the interview.

One traditional Fermi question is this: *How many piano tuners are there in Bristol?*, though that's unlikely to engage most teenagers. A question more likely to interest them would be something like this: *In the great Live Aid rock concert at Wembley Stadium, roughly how many people stood on the pitch/flat area (as opposed to sitting in the stands)?*

The thinking might go something like this (remember, no looking up of data is allowed, it all has to be intelligent guesswork):

1) How big is a football pitch? Call it 100 metres long and 50 metres wide. So its area is $50 \times 100 = 5000$ square metres.
2) How close to each other do people stand in a packed crowd? If they were packed in shoulder to shoulder then maybe two people per metre across and three people per metre front to back. That makes about six people per square metre.
3) That means there's capacity for perhaps 30 000 to stand up on a football pitch, though it would be quite claustrophobic.
4) Football stadiums usually have more flat area than just the pitch, Wembley stadium had space for a running track around the outside. That gives enough space for our 30 000 to spread out a little bit, so stick with that number as an estimate.

(The total attendance at Live Aid was apparently 72 000, so that figure of 30 000 seems reasonable. It's certainly closer than 3000 or 300 000!)

There are hundreds of examples of Fermi questions to tackle, easily found with a quick search on the Web. Here are some that teenagers tend to enjoy:

1) How many times does a typical teenager say the word 'like' in one day?
2) How many hairs are there on Justin Bieber's head?
3) How many teenagers get to perform on British TV each year (either speaking or singing?)
4) In the next minute, how many people in the UK will sneeze?

You can use *Zequals* to tackle all of them.

ESTIMATION

If you do only three things:

- Remember that in everyday life, approximate answers can be as important as precise ones. To be able to work out approximate answers, your teenager needs to know (or be able to mentally figure out) their tables, and how to multiply and divide by tens, hundreds and thousands.
- Look for opportunities to practise estimation with your teenager. Use idle moments (when driving or queuing, for example) to talk about estimating crowd sizes, shop revenues and so on.
- Practise 'Zequals' with your teenager until it is second nature for you both.

RATIOS AND PERCENTAGES

We are offering a great energy deal: 5% off your gas bill and 5% off your electricity bill – which makes 10% off your combined bill.*

The energy salesman quoted above had the misfortune to choose one of the authors as his victim. He therefore found himself having to undergo a ten-minute maths lesson to demonstrate why deducting 5% from two different bills means that the combined bill will also be reduced by 5%. He left looking suitably chastened, and yet there was nothing to stop him attempting the same line on other less mathematically confident customers. And without a strong understanding of percentages, this sort of con trick is extremely easy to miss.

* Door-to-door salesman of a well-known energy company.

Ratios and Proportions

We'll come back to percentages shortly, but they and fractions are part of a wider area of essential maths that we'll call ratios.

The reason we've highlighted ratios as being such a critical skill is that they crop up everywhere. There is hardly an area of practical (or more abstract) maths where ratios are not important. For example:

- When cooking how do you scale up a recipe for four people so that you can serve six people?
- When a builder looks at an architect's plans, how does he work out the length of the marked-out walls?
- How do you work out the VAT that has been charged on a new TV?
- Is a '3 for 2' deal better than a 40% discount at a supermarket?*
- When travelling abroad, how do you know how much cash to take with you?

To answer any of these, and countless other everyday questions, requires an understanding of ratios in one form or another. And ratios are an area in which many teenagers are extremely shaky, because they have never had solid foundations.

The calculation triangle

The explanation that follows might appear so basic that it should be obvious, yet remarkable as it may seem, we think that most teenagers lack this understanding.

* No, a 40% discount is better.'3 for 2' only represents a 33% discount, and it forces you to buy three items when you might only have wanted to buy one.

Learning how to manipulate ratios and fractions starts with a simple relationship between three numbers that can be set out in a triangle. We'll use the example $36 \div 4 = 9$. This can be rewritten in two ways: $36 \div 9 = 4$ and $4 \times 9 = 36$, and we can present this as a triangle:

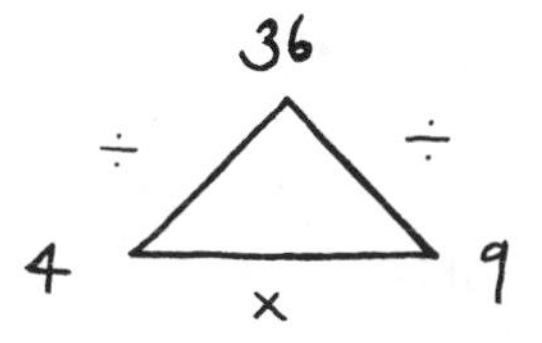

Cover up any corner of the triangle and you can work out the covered number by calculating from the other two.

Any multiplication involving three numbers can be written in a triangle this way, and if we call those numbers A, B and C, then:

> If $A \div B = C$, it is always true that
> $A = B \times C$
> and
> $A \div C = B$

Set out as a triangle, it looks like this:

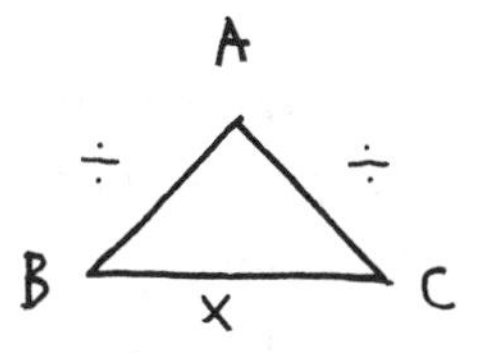

Algebra is the subject of the next chapter, but this particular example involving A, B and C is so important and crops up so often in real-life problems that we've brought it forward to this chapter. We'll come back to it shortly. Your teenager needs to know it, and to know how to apply it in everyday maths situations.

Equivalent fractions and cake

Another simple but important starting point for understanding anything to do with ratios and fractions is the idea that one half is the same as two quarters, and that these are both the same as $3 \div 6$ or $11 \div 22$ or $1000 \div 2000$.

One half is written as $\frac{1}{2}$ – a fraction – but never forget that the horizontal line represents the division symbol. So $\frac{1}{2} = 1 \div 2$.

Also $\frac{1}{2} = 1 \div 2 = 2 \div 4 = 3 \div 6 \ldots$ and these are equal to *any* number divided by its double.

Some people find this even easier to understand with an illustration of a cake. In each of these examples, the amount of cake being taken is the same:

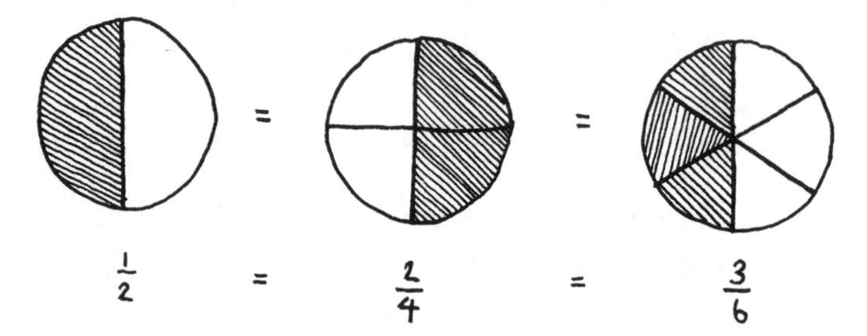

The same principle applies to any fractions, where (if you carry on with the simple cake analogy) the bottom part of the fraction represents the number of slices that the cake has been divided into and the top represents how many of those slices that you take.

So: $5 \div 6$, or $\frac{5}{6}$, means (in cake-speak) five slices of a cake that has been divided into six equal pieces. Incidentally, this is the same amount of cake as you get if you share five cakes equally between six people.

And $\frac{5}{6}$ represents the same amount of cake as $10 \div 12$ of a cake or $40 \div 48$ of a cake. (After a while it is of course important to be able to drop the 'of a cake' phrase, and to be comfortable saying '$\frac{5}{6} = \frac{10}{12}$'.)

What is not so easy, however, is understanding what $\frac{5}{6}$ is equivalent to if you were to divide the cake into, say, eight slices rather than twelve. Now (as we'll discover shortly) the numbers become a lot messier.

The challenge is to work out the value of '?' when $\frac{5}{6} = ? \div 8$.

Teenagers without a proper understanding of what fractions mean often make the mistake of simply adding the same amount to the top and bottom of the fraction. In this example, since the bottom of the fraction increased by 2 (from 6 to 8), it's common for a teenager to increase the top by 2 as well, so that they will reckon that $5 \div 6$ is the same as $7 \div 8$. You can explain why this can't be true by taking a simpler example of $\frac{1}{2}$ and increase the top and bottom of this fraction by 2. It's clear that $\frac{1}{2}$ is not the same as $\frac{3}{4}$.

Ratios and bookmakers

Another way in which fractions can be represented is in the form of a ratio, using a colon. For example, if five cakes have been shared equally between six boys, then cakes have been allocated to boys in the ratio 5:6. However, it is unusual for ratios to be used this way in school. Normally ratios with colons are used to describe the *relative* proportions in which the 'whole' has been divided up. In the cake example, if five slices have been allocated to boys for every one slice that has gone to a girl, then the cake has been divided in the ratio 5:1 boys to girls, and the boys got $\frac{5}{6}$ of the cake while the girls got $\frac{1}{6}$.

This way of writing ratios to show relative proportions using colons is also used in betting, where a racehorse might be quoted as having odds of 2:1 against. What this means is that the horse is expected to lose two times for every one time that it wins. In other words, odds of 2:1 mean it will win one race out of three, or, as a fraction, it will win $\frac{1}{3}$ of its races. The fact that 2:1 (or 1:2) typically means $\frac{1}{3}$ rather than $\frac{1}{2}$ causes a LOT of confusion with teenagers.

For this reason, we will avoid representing fractions with colons from now on. (If you went to school in Germany, you will know that the colon is used there as the symbol for division. In fact, in Britain we use the colon too, but we put a line through the middle of it, like this ÷.)

Working Out How to Scale Up

Now we come to the crucial part of fractions and ratios, the skill that is vital in solving so many real-life problems. How *do* you fill in the question mark here:

a) $5 \div 6 = ? \div 8$

For that matter, how do you work out the question mark if it's in a different position, for example:

b) $7 \div ? = 21 \div 40$

For question a) $5 \div 6 = ? \div 8$, a calculator will tell you that $5 \div 6 = 0.833$. So if $0.833 = ? \div 8$, then we know that $? = 8 \times 0.833 = 6.667$ (or six and two thirds)

Similarly, for question b) $7 \div ? = 21 \div 40$, in other words $7 \div ? = 0.525$

So: $? = 7 \div 0.525 = 13.333$

Both of these examples involve rearranging the equation in the way we demonstrated in the calculation triangle on page 49.

Most adults would use a calculator in these examples. But for those confident with numbers, mental methods can be just as quick. In the second example, where $\frac{7}{?} = \frac{21}{40}$, if you look at the top numbers on each fraction, you'll spot that 21 is three times 7, so for the two fractions to be equivalent, the bottom number 40 must be three times '?', so $? = \frac{40}{3}$, which is $13\frac{1}{3}$ or 13.333.

If your maths is reasonably strong, then all of this might seem obvious. But it is not obvious, certainly not second nature, to many teenagers, and it needs to be if they are to cope with scaling-up and conversion (whether they go on to become a nurse, a chef, a salesperson, a handyman or an engineer).

TEST YOURSELF

1) What is $\frac{3}{4}$ written as twelfths? (In other words, $\frac{3}{4} = \frac{?}{12}$.)
2) Work out the value of the question mark here: $0.8 = \frac{5}{?}$
3) 6 is to 20 as 15 is to what?
4) If three teabags of a particular brand are enough to make tea for four people, how many people will 80 teabags cater for?

Percentages

The most common way in which ratios crop up in adult life is as percentages. Special offers, pay rises, exam scores, health risks, weather forecasts . . . percentages are everywhere. Not only are they universal, they are also universally misused and misunderstood. Without an understanding of percentages, it's hard to make good financial decisions and very easy to get manipulated. And when it comes to people getting duped by maths, percentages are probably public enemy number one.

- Percentages are used to sensationalise or send scary messages. 'Eating red meat increases your chance of a heart attack by 13%,' said one story typical of many. 13% sounds quite big. But in stories like this, you have to ask '13% of *what*?' It's not 13% of the whole population that are going to have a heart attack, it's 13% of the smaller number who are already destined to have a heart attack. If only 1 in 1000 people are going to have a heart attack, a tiny number, then 13% more than that means that 1.13 in 1000 will have a heart attack: still a tiny number.
- Percentages can make deals sound better than they are. '20% off the main course!' says a restaurant. Yes, but when you factor in the wine, extra vegetables, starter, pudding and coffee, you'll be lucky if the discount on the whole outing is even 10%.
- Percentages get confused with 'percentage points'. When price inflation increases from 1% to 2% that means it has doubled. So if you want to scare the public you could say inflation has gone up by 100%! But the news will report it has increased by 1%, meaning 'one percentage point'. Not nearly as scary sounding, but just as much of a price increase whichever way you present it.

It's easily forgotten that percentages are really just fractions or ratios by another name. All they do is relate any ratios to

100. For example $\frac{3}{4}$ is 75%, which can be written out as: $\frac{3}{4} = 75 \div 100$.

Percentage problems crop up in several different forms:

1) Convert a ratio into a percentage

Want to work out $31 \div 40$ as a percentage? It's just the ABC calculation triangle all over again:

$31 \div 40 = 0.775 = ? \div 100$

So $? = 77.5$

(We say more about converting fractions into percentages on page 216.)

2) Increase or reduce an amount by a percentage

Discounts and taxes typically involve adding or removing a percentage of a particular amount. There are two ways to calculate an increase (or decrease) of, say, 20%.

Let's say you are working out the VAT to add on to the price of something. The slow way is to first calculate 20% of the price, then add that value on to the price. It's handy to remember exactly what '20 per cent of' means: 'per' means divided by, 'cent' means 100 and 'of' in this case is the same as 'multiplied by'. So 20% of £500 in longhand is $(20 \div 100) \times 500 = £100$.

The quicker way of increasing a price by 20% is simply to multiply the price by $(100 + 20) \div 100$, or 1.20. Reducing the price by 20% means multiplying the price by $(100 - 20) \div 100$, or 0.8.

3) Work out the original amount before it was adjusted by a percentage

This crops up less often in everyday life than the other examples, but it's still important, and it's the percentage calculation that people most often struggle with. Let's say you buy a TV for £300, and that price includes VAT at 20%. How much was the VAT? The pitfall that teenagers fall into all the time is to work out 20% of the final amount, in this case 20% of £300 (= £60) and just subtract it. In other words, they say that the price was £240 and the VAT was £60. Although the figures add to £300, the VAT figure of £60 is too high – 20% of £240 is £48.

Three logical steps are needed instead of one:

1) The original price $\times$ 1.20 (i.e., including the VAT) = £300
2) Therefore the original price must be £300 $\div$ 1.2 = £250
3) The VAT is therefore £300 – £250 = £50

Tip: calculating percentages

Many people (adults as well as teenagers) struggle to calculate percentages without a calculator. What's 30% of 90? It's hard! When calculating an awkward percentage, always start by working out what 10% is. 10% of 90 is 9 – easy. Then, to work out a different percentage, simply scale the 10% up or down to that value. Since 10% of 90 is 9, then 30% will be three times as large, $3 \times 9 = 27$. What about 5% of 90? 5% is half of 10%, and $\frac{1}{2} \times 9 = 4.5$.

TEST YOURSELF

1) What is 40% of 350?
2) One of the authors was in an off-licence that was offering 20% off a dozen bottles of wine. The assistant said he only knew how to take off 10% so he did that, and then took off another 10% off the lower bill. Only on the way home did it become clear why this was not the same deal. Who was diddled?

3) Kate and Jasmine both earn the same wage. Kate's wage is cut by 10% and then the next month she gets a 20% increase. Meanwhile Jasmine's wage is increased by 20% and the next month she gets a 10% pay cut. After these changes who has the highest wage? (a) Kate (b) Jasmine or (c) At the end they are both on the same wage.

RATIOS

If you do only three things . . .

- Make sure your teenager knows the ABC calculation triangle, using real number examples (such as $36 = 4 \times 9$).
- Reinforce the message that ratios, scaling-up, fractions and percentages are all just different ways of saying the same thing – and they crop up everywhere in maths.
- The easiest way to work out percentages is usually to calculate 10% and then scale up or down from there. To work out 30%, first work out 10% and then multiply the answer by three.

PROBLEMS AND PUZZLES

Here is a selection of problems and puzzles that between them bring out many of the important features of problem-solving. None of these problems requires any advanced maths, but that doesn't mean that they are easy. In fact, a couple of them have been known to take people several hours to solve, and there are some that you might not solve at all.

If you try one of these problems with your teenager, resist the temptation to look up the answer in the back. The feeling of being 'stuck' is an important part of problem-solving, and one that your teenager needs to experience and to become comfortable with if they are to be confident problem-solvers later on. If you really can't make any progress, there are some hints to help you.

Who cares?

In one or two cases, it's quite possible that you will look at the problem and find yourself asking the question 'Who cares?' And it is a perfectly legitimate question. Nobody solves problems unless they are motivated to do so. Your teenager will only learn about problem-solving if they are given problems that they care about solving. The motivation for solving problems might come from the problem being intrinsically interesting, or from a growing curiosity when they discover things are not as obvious as they first appeared to be. Sometimes movitation comes from the prospect of winning a prize, or simply from competing with others to see who can find the solution first.

1) THE GAZELLE PUZZLE

Lay five matchsticks on a table to make a gazelle, like this. At the moment you are looking at an image of his left side. Can you transform this into an image of the gazelle's *right* side, by moving:

a) Exactly two matches
b) Exactly one match?

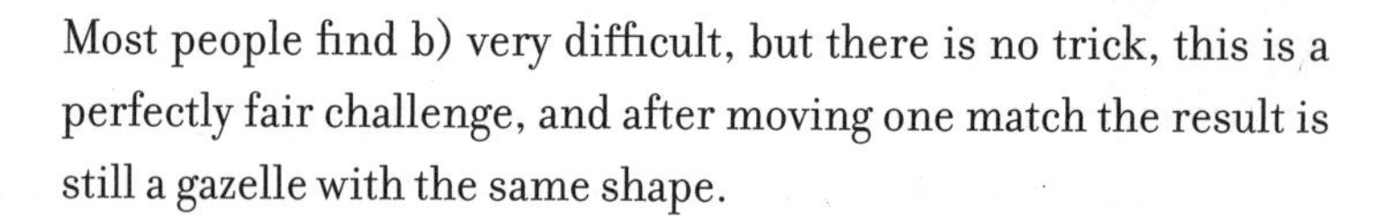

Most people find b) very difficult, but there is no trick, this is a perfectly fair challenge, and after moving one match the result is still a gazelle with the same shape.

Hint Which match do you need to remove in order to leave a symmetrical pattern of matches?

2) CLOCK HANDS

Between noon and midnight, how many times are the hands of a clock at right angles to each other? (3 p.m. is one example of when it happens – but it's not the only one.)

Hint The first time the hands will be at 90 degrees isn't quite quarter past twelve, and about half an hour later the hands are at 90 degrees again.

3) THE DICTIONARIES

There is an illustrated dictionary sitting on a bookshelf. The dictionary has two volumes, the A–L volume on the left and the M–Z volume on the right. Each volume is roughly 5cm thick. There is a bookmark sticking out of the first volume to indicate 'Aardvark' and another bookmark in the second volume indicating 'Zoo'. How far apart are the two bookmarks? (Remember the Bow and Arrow question on page 5 – do not trust the first answer that you come up with, because in this case it is very likely to be wrong. There is no trick.)

Hint Check with a real dictionary on the shelf: whereabouts is the first page, on the left or the right?

4) THE BABYSITTER

You have agreed to babysit for three hours, and it's agreed you will be paid £20. Unfortunately your bus is delayed, so you ring to say that you are going to be half an hour late, which means you will only be able to babysit for $2\frac{1}{2}$ hours. How much should you now be paid?

Hint No hints, but what do you think the issues are?

5) DATE PUZZLE

Brian has a simple calendar on his desk. It has two cubes, and both cubes have a different digit between 0 and 9 on each face. Brian uses the cubes to form all the dates between the 1st and the 31st of each month. For example, on the fifth of the month he turns one cube to show '0' and the other to show '5', while on the twenty-third he turns one cube to '2' and the other to '3'. He always uses both cubes. What numbers are on each cube?

On the 7th of the month the two cubes are placed like this.

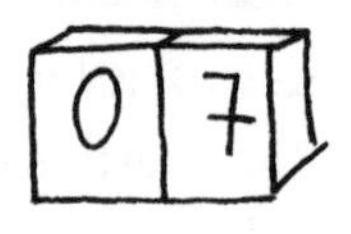

Hint To make 11 and 22 you need 1 and 2 on both cubes. And you also need to make 01, 02, 03 . . . all the way up to 09 which means you'll need 0 on both cubes. It seems there is one digit too many to squeeze on to one of the cubes.

6) TWO PLUGHOLES

There are two plugholes in Sam's bath. (Sam is a maths teacher, and it's a well-known fact that maths teachers' baths always have two plugholes.) If the bath is full and she pulls out the first plug, it takes six minutes to empty the bath. On the other hand, if the bath is full and she pulls out just the second plug, it takes only four minutes for the bath to empty. If the bath is full and she pulls out both plugs, how long does the bath take to empty?

Hint The water passes through one plughole at the rate of $\frac{1}{6}$ of a bath per minute, and the other one at a rate of $\frac{1}{4}$ of a bath per minute. So after one minute how much water has left the bath if both plugs are removed?

7) HOW MANY SQUARES?

On an ordinary 8×8 chessboard there are 64 small squares. But there are other squares too, for example the large square that forms the outside of the board. How many squares are there in total on a chessboard?

Hint Start by thinking about a simpler problem. How many squares are there on a 2×2 chessboard? And on a 3×3 chessboard?

8) ESCAPING PRISONERS

In the high-security Dun-Thievin prison there are 100 prisoners, each in his own cell, and 100 prison guards. One night, with all the cells locked, the prison guards have a party and end up rather the worse for wear. The first prison guard, forgetting that all the cells are locked, visits each one and turns the key, thus unlocking every cell. The next guard, for reasons best known to himself, visits every second cell (numbers 2, 4, 6, 8 and so on up to 100) turning the keys in those cells. The third guard visits cells 3, 6, 9 and so on up to 99, turning the keys in those cells, locking some of them and unlocking others. The same pattern happens with each guard, the fourth going to cells 4, 8, 12, the fifth going to 5, 10, 15, etc, all the way up to guard number 100 who simply turns the key in cell 100. All the guards now go to bed. Which prisoners can now escape?

Hint How many prison guards visit cell number 4? And how many visit cell number 6? Which of these ends up unlocked?

9) THE DIE-HARD JUGS PUZZLE

This puzzle featured in the film *Die Hard with a Vengeance,* when Bruce Willis and Samuel L. Jackson were given five minutes to solve it. Most people take rather longer! You are given two water containers, one that can hold exactly five gallons and one that holds three gallons. You have an unlimited supply of water, and need to put exactly four gallons

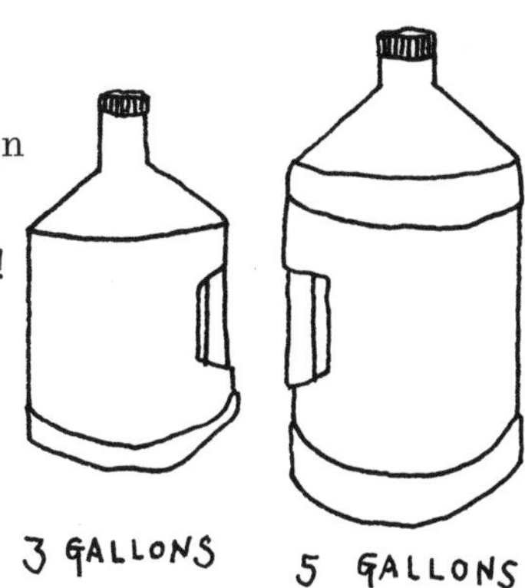

into the larger container and set it on a weighing machine. If you are out by more than a couple of teaspoons then you and your colleagues will meet a gruesome end. How do you do it?

Hint What happens if you empty the full five-gallon jug into the empty three-gallon jug?

10) THE STATUE IN THE DESERT

On your journey through a distant desert land, you come upon a mysterious tower in the middle of a flat plain. Keen to report back accurately on your discovery, you want to know exactly how tall this statue is. The statue looks like it's the height of a multi-storey building, but its surface is extremely smooth and slippery, and impossible to climb. You have in your rucksack the following equipment: a three-metre tape measure, a mobile phone that's about to lose its power (maybe thirty seconds left), some string, some plasticine, a spare T-shirt, a large bottle of water, a pack of sandwiches held together with an elastic band, a floppy hat, a pencil, a hardback book called *Insects of the Desert* and a pair of binoculars. How would you work out the height of the statue?

Hint There isn't any one right answer to this, though some answers are a lot more practical than others. In real life you aren't expected to use all of the equipment that you have at your disposal.

PART THREE
THE MATHS

ALGEBRA

Teacher: So now class we need to find x.

Pupil: Sir, I thought we found it yesterday.

What *is* Algebra?

We've put algebra first not because it's the first maths that teenagers encounter, but because it is overwhelmingly the part of school maths that teenagers and parents struggle to get to grips with.

Algebra is the part of maths where numbers get replaced by letters or other symbols. Or at least that's the very simplistic way of explaining it. The truth is that it is the foundation of a huge amount of maths, since it is an essential tool in understanding everything from probability to mechanics. The word algebra – or 'al-jabr' in its original Arabic form – was coined over a thousand years ago, meaning 'putting back together'. Many of the ideas of algebra go back even further, to the Babylonians around 2000 BC.

Could modern society exist without algebra? Almost certainly not, because the alternative – trying to solve complicated problems using trial and error, or by drawing pictures – is far too slow and cumbersome, and sometimes it is simply not accurate enough.

Unfortunately, it is also true that algebra was the point at which many adults remember losing touch with mathematics. And many teenagers feel the same way. In our straw poll of teenagers across the UK, it would seem that algebra is the part of maths that gives them the most problems.

There are several reasons why algebra is such a barrier. One is that it appears to be such an abstract topic, manipulating all sorts of symbols that don't seem to represent anything in the real world. Another is that unless you are comfortable with the foundations that lead up to it, algebra can look like an alien code. Imagine opening a maths book and encountering an equation such as $3x^2 + 4y^2 = 24$. Out of context, it's only reasonable to ask: 'What does this mean? What am I supposed to do with it? And who cares?'

We've divided algebra into three chapters. The first addresses the question of what it is and what it is for, the others look at how to solve algebra problems and how graphs relate to algebra and to the real world.

WHAT IS x?

Think of a number, any number you want. Now double it. Add 12. Then divide your answer by 2. Finally, take away the number you first thought of.

And as the drums roll we reveal that you finished on the number . . . 6. (If you didn't, you made an arithmetical mistake somewhere, check your answers again.)

How amazing that this little trick works whatever number you start with, no matter how big or small. It works for fractions, too. Think of a number . . . '$\frac{1}{2}$' . . . double it '1' . . . add twelve . . . '13' . . . Halve it '$6\frac{1}{2}$' . . . take away the number you first thought of . . . 'still 6!'.

It even works for negative numbers, as long as you remember the rules for taking away a negative (see page 199). Start with -2, double to make -4, add twelve gives $+8$, halve it gives 4, then take away -2 (which is the same as adding $+2$) and you get 6. It's almost magic.

Think-of-a-number tricks have a long history. Even Leonardo da Vinci played with them. There is an example where Leonardo asks his readers to select any number of beans in one hand, and after various manipulations, the result is always 13 beans.

Apart from being a little light entertainment, it turns out that these tricks are also a great introduction to algebra.

In our example, why is the final answer always 6?

To see why, imagine that the number that you first thought applies to some marbles, and you have put these marbles in a box. We'll use a symbol for the box that contains the marbles: ❒

Now do the trick on ❒:
Think of a number: ❒.
Double it: you get ❒❒.
Add twelve: you have ❒❒ *plus* another 12 marbles. (*Not* 12 boxes, you haven't added 12 boxes, just 12 marbles.)
Halve it: half of 2 boxes plus 12 is ❒ + 6.
Take away the number you first thought of: ❒ + 6 − ❒ = 6.

This trick is all about getting rid of the number that was in the box, and it doesn't matter what that number was because it disappears at the end in any case.

At its heart, getting rid of the box is what algebra is all about. The biggest difference between this example of algebra and the stuff that your teenager has to do is that at school, the number that we called ❒ is referred to as 'x'. And that's where the trouble really begins.

The story of *X*

These days it is standard practice among mathematicians to call an unknown number 'x', something we all take for granted because that's what we were taught at school. But of course somebody had to invent this idea. Four thousand years ago, the following question was posed in a set of Egyptian maths problems now known as the Rhind Papyrus: *A quantity and its one-seventh added together become 19. What is the quantity?*

The reason why this is so long-winded is that the Egyptians had no shorthand code for saying 'a quantity'. Today we would write this problem much more succinctly: *Solve:* $x + \frac{x}{7} = 19$.

It took hundreds of years for mathematicians to figure out how to write down mathematical problems in a concise way that everybody would understand. It was about five hundred years ago that Rene Descartes (most famous for his quote 'I think therefore I am') decided to formally declare that numbers that were unknown or could have any value (*variables*) should always be known as x, y and z, whereas numbers that were constant should be known as a, b and c.

That convention has stuck ever since, though unfortunately it has made life unnecessarily complicated for maths pupils because of the resemblance of x to the multiplication sign $\times$. Writing 3×7 doesn't cause confusion, but if you want to multiply x and y together, then $x \times y$ can easily be misread. For this reason, the multiplication symbol in algebra is replaced with a dot ($x.y$) or, more commonly, it is left out altogether with all spaces removed. So in a mathematical expression, xy means x multiplied by y. And it's all Descartes' fault.

Use Doris to make algebra friendlier

Many teenagers go cold when they encounter questions involving x; in some cases their minds even go blank. This is mainly because they associate it with solving difficult maths questions. When encountering x in a problem, it sometimes pays to lighten the mood by renaming x as something more human, like Doris. It doesn't make the problem any easier, but sometimes taking the serious tone out of the question can make it less intimidating. Instead of solving $x + \frac{x}{7} = 19$, try Doris + (Doris ÷ 7) = 19.

What is *X* for?

So what is algebra actually *for*? We've already seen that it is a concise way of saying 'the quantity that we do not know'. But why do we want to know that quantity in the first place? There are actually two different reasons for involving x, which represent the two reasons for algebra.

1) To help you to solve problems

Algebra can be an extremely useful tool in helping you to solve mathematical problems, by allowing you to use symbols to represent quantities that you want to find out but don't yet know. These may be real problems, such as figuring out what your mobile phone bill might be in the next six months, or 'made up' problems and puzzles, which are there to help you practise your skills or even just to have fun (some people really do get their kicks from solving algebra puzzles). Here is an example of such a puzzle:

A woman is five times as old as her daughter. In six years' time she will be three times as old as her daughter is then. How old is the woman now?

Puzzles like this are designed to seem confusing, rather like verbal tongue-twisters, and simple algebra is one of the best ways to untangle the mystery. If we call the age of the woman W and that of the daughter D, then in this case we know that:

a) The woman's age W is currently five times that of her daughter, i.e. $\underline{W = 5D}$
b) In six years' time (when the woman will be W + 6, and the daughter will be D + 6) the woman's age will be three times that of the daughter. In other words: $\underline{W + 6 = 3(D + 6)}$.

In just a handful of symbols, we've been able to set out the whole problem. Solving it is another matter (for the record, these are what is known as *simultaneous equations* and the daughter is 6 while the woman is 30). But the main point here is that one of the secrets of problem-solving is being able to express a problem accurately in the first place. This is where algebra can be an invaluable aid.

2) To describe the rules and patterns of maths

Algebra is an extremely concise way of expressing rules about the way mathematics works. If you want to get lyrical about it, algebra helps to unlock the secrets of the universe. Here is a trivial example of some maths that is always true and what is known as the commutative law: *The order in which you add numbers up makes no difference (so, for example, $6 + 13 = 13 + 6$).*

It took a whole sentence to explain that. It is so much neater to be able to make exactly the same point using just a few symbols: $x + y = y + x$, *for all* x *and* y.

Now let's look at a less obvious and therefore rather more interesting example. Here is a mathematical idea expressed in plain English. Can you get your head around it?

> *If you take any two numbers, then the sum of those two numbers multiplied by their difference is always equal to the difference between the squares of those two numbers. Take for example the two numbers 5 and 4. The rule says that their sum* $(5 + 4 = 9)$ *multiplied by their difference* $(5 - 4 = 1)$ *will always be the difference between the squares of those numbers, so for example:* $5^2 - 4^2 = 25 - 16 = 9$*.*

Did you follow that? Unless you do a lot of maths, your eyes probably glazed over and you no doubt had to reread it to be sure that you had grasped the idea. Writing out mathematical ideas longhand can be mind-bending, just like the earlier puzzle about the age of the woman and her daughter.

If instead we call those two numbers a and b then we can express the whole idea algebraically like this:

$(a + b)(a - b) = a^2 - b^2$, *for all* a *and* b.

Once you're comfortable with this language of maths, then there is little doubt that it's actually far easier to get your head around a rule using the language of algebra rather than spelling it out in long sentences.

There is, needless to say, rather more to algebra than that. For a start, algebra can also be a way to make discoveries about maths that were not known before. But up to GCSE most teenagers in most circumstances will be using algebra for one of the two reasons above.

Confusion Between the Two Uses of X

We've just seen the two uses of algebra are for problem-solving and for finding general rules. Because these uses are very different, they are the source of one of the main confusions that teenagers have with algebra, which can be boiled down to the question: *What does 'x' actually mean?*

When you are solving a problem, x (or whichever symbol is being used) is used to represent the number you are trying to find. For obvious reasons, x here is called an *unknown*.

However, in situations where you are merely demonstrating the connections between different mathematical expressions, x represents any number at all, and is called a *variable*.

If you are presented with this:

$$3x + 2 = 11$$

then it makes sense to be asked to 'Find x'. Here, x is the unknown (and in case you haven't worked it out already, $x = 3$).

But presented with this:

$$y = 3x + 2$$

it makes no sense to 'Find x', since here x is a variable and can be any value. If you are confident and experienced in maths then that is probably obvious, but if you aren't, then you are in the same situation as most teenagers, who would see little or no difference between the two equations above.

There are two ways to get comfortable with seeing that x can be used in two different ways.

The first way is by setting x in a real-life situation. For example, x might represent the amount of time you spend talking on your mobile phone. Imagine if the cost of a premium overseas call on your phone is £1.25 + £0.15 (i.e. 15p) per minute.

The cost of a call, C, is therefore $C = 1.25 + 0.15x$, where x is the call duration in minutes. You don't need to think twice that x is a variable here.

But if you are told that a call cost £4.25, and want to know how long it took, then it's obvious that in this example there is an unknown that you want to calculate. In this case $4.25 = 1.25 + 0.15x$, and so (as you can check, or work out) $x = 20$ minutes.

The second way of overcoming confusion about what x means is simply through practice, so that after a while it becomes second nature.

Spotting the difference between an unknown and a variable

Here's an easy way to spot the two types of equation:*

- If an equation contains only one letter (even if that letter appears more than once, for example $3x + 7 = 2x$) then x is an unknown that you are expected to find.
- If there are at least two different letters, all you can find is the simplest way of expressing one letter in terms of the other. For example $y + 3x = 2y + 1$ is most simply expressed as $3x = y + 1$, but we don't know what x is unless we also know what y is. We explain more about rearranging equations like this in the next chapter.

* Mathematicians will point out that there are *special situations* where these rules don't work.

Equations, Expressions and Formulae

It's common to hear teenagers (and indeed adults) describing any combination of mathematical symbols and numbers as an 'equation', but an equation has a particular meaning, and is very easy to spot. An equation is a statement that contains the EQUALS symbol, and which states that something equals something else.

So, for example, these are equations:

a) $17 = 9 + 8$
b) Total eggs = No. of boxes multiplied by 6*
c) $2x + 3 = x^2 \div 7$

Replace the '=' symbol with '>' (greater than) or '<' (less than) and instead of an equation you have an *inequality*. And if there are none of these symbols present, then what you have is simply known as an *expression*. So, for example, $2k + 6$ is an expression.

A *formula* is a particular type of equation. A formula is a general rule for working out something by feeding numbers into the other side of the equation. For example, if you buy eggs in boxes of half a dozen, then the *formula* for working out how many eggs you have is 'the number of boxes × 6'.

The formula for working out children's ages

Do you struggle to remember what age a child is when you hear they are in 'Year 3' or 'Year 10' (this 'new' terminology was actually introduced in 1989 as part of the National Curriculum, and replaced the old terms 'third form', 'fifth form' and so on – though confusingly Years 12 and 13 are still widely referred to as

* The standard egg box contains six eggs, of course.

'the sixth form'). There's a simple formula for knowing the age of children when they start a particular year at school:

Age = Year + 4

So as a rule, all Year 11 children start the school year at the age of fifteen, and as the terms go by they all have their sixteenth birthdays (with all the mayhem and parental anxiety that those entail).

TEST YOURSELF

Formula, equation, both or neither?

a) $F = \frac{9}{5}C + 32$
b) Sally's age = three times Bobby's age + 4
c) πr^2
d) $x + 3 = x + 4$

N and the nth Term

In algebra there is another letter, n, that crops up almost as often as x. Typically, n crops up in problems like: *The nth term of a sequence is n + 9. Work out the first four terms.*

In a newspaper article the journalist Lucy Mangan described her reaction to this particular example: *I could look at this sentence until the end of time and never be able to work out what it means.*

To be honest, the nth term example is far less daunting than challenges such as 'Factorise $x^2 + 3x + 2$' (see the next chapter), but many teenagers, and their parents, do find the idea of the 'nth term' difficult. It's not just the word *nth*, but also the word *term*, a familiar everyday word that has a different, very particular, meaning in maths. In maths, if you have a series of numbers or expressions, each item in the series is called a term.

But what's the point of finding the nth term in the first place? The answer is that this is how algebra helps us express our natural human urge to find patterns. It starts with simple things like predicting the next house number along a street '21, 23, 25 . . . so house number 31 is going to be another three houses from here' – and goes all the way to the brilliant work done by cryptographers at Bletchley Park who were able to decipher the sophisticated patterns hidden within the Enigma machines.

Why is it always 'n' rather than some other letter? By convention, and remember it *is* only a convention, 'n' is used to represent whole, counting numbers. So you might refer to *n* cars on the road, or *n* people in a house, because these are things that you count, but you wouldn't use *n* for the height of a tree or the weight of a suitcase.

It's an appropriate letter to use. It is, after all, the initial letter of '**N**umber', and you can think of it as representing a**n**y counting number you want. It has even acquired its own word, 'nth' (pronounced 'enth' of course), which sometimes enters everyday conversation: 'So I told him for the nth time to get lost.'

Usually **n** is used to describe the pattern in a sequence of numbers, In the sequence 2, 4, 6, 8 . . . for example, the first term is 2, the second term is 4, and the nth term is simply 2n.

If there's ever any confusion about this, by far the best thing to do is to test out the rule using actual numbers. For example, in the problem that started this section: *The nth term of a sequence is* $n + 9$, replace *n* with 1, 2, 3 and so on. The first term of a sequence is 1 plus 9, the second term of a sequence is 2 plus 9, the third term of a sequence is 3 plus 9, and so on.

Incidentally, n can be written as a capital or lower-case letter, though capitals and lower case tend to be used for different purposes. Lower-case n usually refers to any number in a sequence, while capital N means the total number.

The nth term: think of pigeonholes

One way of demystifying 'nth terms' is to imagine a line of pigeonholes, numbered 1, 2, 3 . . . Whatever goes into the first pigeonhole is the first term, the second pigeonhole is the second term and so on.

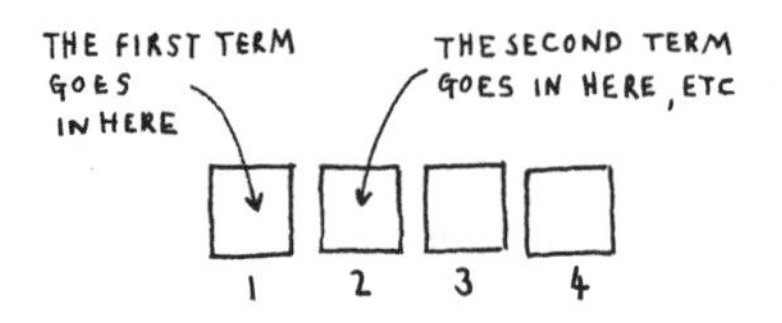

When you are told that (for example) 'the nth term is $5n + 7$', then 'n' means 'the number of the pigeonhole', and '$5n + 7$' is the rule for working out what goes into that pigeonhole. In this case, to find what goes in each pigeonhole the rule is to multiply the number of the pigeonhole by 5, then add 7. So the term that goes in the first pigeonhole is $5 \times 1 + 7$ (= 12), the second pigeonhole is $5 \times 2 + 7$ (= 17), the third is $5 \times 3 + 7$ (= 22) and so on. Following the rule '$5n + 7$' the series begins 12, 17, 22, 27 . . .

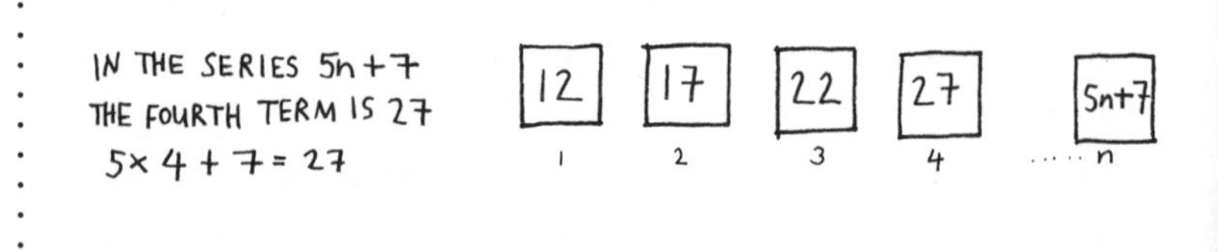

TEST YOURSELF

a) The nth term in a sequence is $2n + 5$. What is the value of the 5th term? And the 105th term?

b) The nth term in a sequence is $4n + 3(n-1) + 2$. What is the value of the 10th term?

c) A sequence of numbers is calculated as

1st	$(7 \times 1) + (3 \times 2) + 5$
2nd	$(7 \times 2) + (3 \times 3) + 5$
3rd	$(7 \times 3) + (3 \times 4) + 5$
4th	$(7 \times 4) + (3 \times 5) + 5$

and so on. Write down an expression for the nth term.

Real Life into Algebra

Turning a real-life situation into algebra is not always as easy as it looks. Imagine you are a barman mixing a gin and tonic. Suppose G represents the amount of gin, and T represents the amount of tonic, and you've been told that the rule in this bar is: *for each single unit of gin there should be two units of tonic.*

What is the equation that connects the amount of gin to the amount of tonic ?

The vast majority of people posed this problem will write down something like this:

$$G = 2T$$

because that looks like the way you'd write 'there is one gin for every two tonics'.

But does it make sense? Suppose you have one unit of tonic. How much gin should you add? If the formula $G = 2T$ is right, the amount of gin you need is two times the amount of tonic, as the illustration shows:

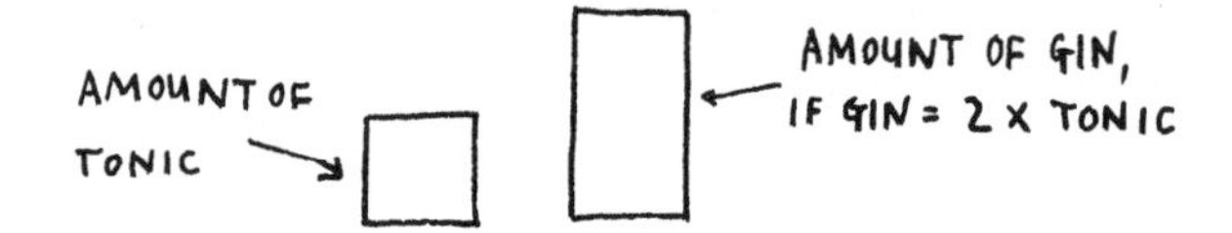

So you'd put one tonic into a glass and two gins. Try it – a G&T that strong will have you under the table in no time.

The key thing here is that the statement 'one gin for every two tonics' actually means 'the number of tonics is twice the number of gins', so the correct formula for a G&T is actually: $T = 2G$.

This shows just how careful you have to be when converting everyday phrases into mathematical language, and of all the stumbling blocks in algebra, it is probably this one that is the most significant – though teenagers tend to be protected from it, since they are rarely asked to turn a problem into algebra before solving it.

There's a general rule for dealing with this. Whenever you try to turn a word problem into a mathematical expression, test it out by putting some real numbers into it to see that it makes sense.

TEST YOURSELF

P is the number of professors at a university and S the number of students. Every professor supervises six students. What is the formula that links P and S?

Another Think-Of-A-Number Trick

We started the chapter with a think-of-a-number trick. Here's a more sophisticated one (variants of which occasionally become viral on Facebook). For this example, a calculator will help.

How many burgers did you eat last week?*
Multiply that number by 2.
Add 5 to that answer.
Multiply by 50.
Add the current year (so if you are reading this in the year this book was first published, add 2013).
Subtract 251.
If you have already had a birthday this year, add 1 for luck.
Subtract the year that you were born.

The number you've ended up with is *the number of burgers you ate, followed by your age*.

How does it work?

Suppose the original number of burgers that you ate was B. Let's use that to track through the calculations.

How many burgers?	B
Multiply that number by 2:	$2B$
Add five to that answer:	$2B + 5$
Multiply by 50:	$100B + 250$
Add the current year (e.g. 2013):	$100B + 250 + 2013$
Subtract 251:	$100B + 2012$
Add 1 if you've had a birthday:	$100B + 2013$

* Instead of burgers, you can make this the number of cousins you have, or any other number that might sensibly be in the range of 1 to 10.

Subtract the year that you were born. $100B + 2013 -$ birth year

What does this mean? The number in the hundreds column will be your 'burger number'. And since 2013 minus your birth year was your age in the year 2013, the last two digits must be your age.

WHAT IS x?

If you do only three things . . .

- Remind your teenager that algebra has two important but very different uses, both of which might involve 'x': 1) as a shorthand way of solving mathematical problems (where x is an unknown to be worked out), and 2) as a way of describing patterns (where x is a variable that can have any value).
- Many teenagers are uncomfortable with the use of x and y. You can humanise variables by giving them any name you like, such as Doris.
- When dealing with an algebraic expression, it helps to replace the letters with numbers to get a sense of what is going on.

MANIPULATING AND SOLVING ALGEBRA

'Simplify', 'Factorise', 'Expand', 'Solve for x', 'Rearrange to make y the subject' and 'Substitute for x'.

That little collection of mathematical terminology is guaranteed to send a shiver down the spines of countless teenagers and their parents.

For most people faced with a word like 'factorise', it's not just a question of 'how', but also, probably more than any area of secondary school maths, '*why*?'. The explanation of this latter question can be particularly challenging. After all, there are very few adults who have ever had a need to 'factorise' or 'expand' any algebra other than in a maths exam.

What might surprise you, therefore, is that there's a small but significant proportion of teenagers who, when asked which is their favourite bit of maths, quote the manipulation and solving of algebra at the top of the list. Why? Because they think of it as solving a puzzle. It's all about thinking, and requires only a minimal amount of calculation. (And if there's one thing many teens hate in maths, it's grubby calculations with big numbers.)

However, there must be more to it than solving puzzles, and the truth is that 'simplify', 'rearrange' and 'factorise' make up an essential toolkit that is as important to tackling serious maths problems as knowing how to make a white sauce is for a chef. Like white sauce, these techniques aren't that interesting or useful in themselves, but without them it would be impossible to get beyond the basics of the subject.

Part of the problem with much of this mathematical terminology is that the meaning is far from clear. The instruction 'expand', for example, bears little relation to the everyday use of the word. This reminds us of the example of a pupil's deliberate misinterpretation of the word expand, which has been circulating on the Internet for years:

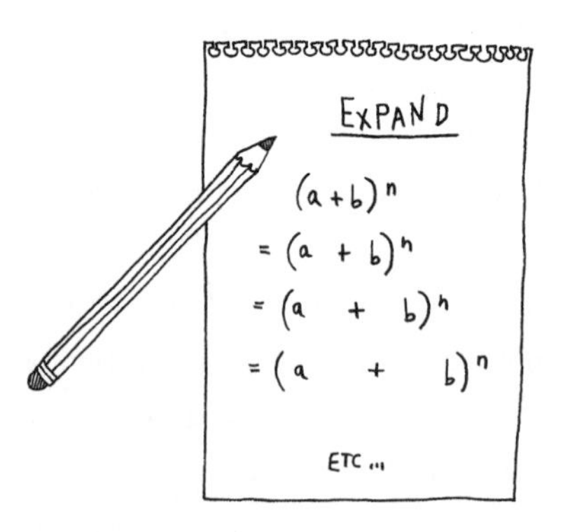

In the same way, 'simplify' doesn't always end up with something that (to an untrained eye) looks 'simpler', and 'solving' doesn't always end up with a specific 'solution' as an answer.

Once you accept that you have entered a world of specialist mathematician-speak with its own private code, you can (at least partly) relax.

The Grid Method – Algebra and Milk Crates

When your child was in primary school, they were probably taught the so-called *grid method* of multiplication. This method, which we'll explain in a moment, is taught for two reasons. First, it's an easier method to understand than the traditional 'black box' method of long multiplication. And second (though they won't have been told this at the time), it is a good foundation for learning one of the most important tools of algebra.

A simple way to understand the grid method, and its connection with algebra, is to imagine counting the bottles in crates of milk. We will introduce it in two stages.

Step 1: Here is a milk crate. How many bottles are there?

You don't have to think twice, it is four rows of five, of course, and $4 \times 5 = 20$.

Or perhaps you spotted that it's five columns of four, and 5×4 is also 20. The important idea here is that any multiplication of whole numbers can be thought of as multiplying the rows and columns of bottles in a milk crate, and when thought of this way, it is obvious that *the order in which you multiply two numbers* does not matter: $5 \times 4 = 4 \times 5 = 20$. (In fact, the order in which you multiply *any* group of numbers does not matter: $3 \times 7 \times 11 \times 13 \times 37$ is the same as $7 \times 11 \times 37 \times 3 \times 13$, or any other ordering. If you happen to have a calculator to hand you might like to test it out with these five numbers – you should get a surprisingly pretty answer, in whichever order you enter the numbers.)

Step 2: Now imagine a huge milk crate, 17 rows by 23 columns.

23 COLUMNS

17 ROWS

We know that the total number of bottles is $17 \times 23 = 23 \times 17$.

How do we work out the answer? One simple way is to divide the grid up into rectangular blocks. The 17 rows can be divided into 10 + 7, for example, while the columns can be split up as 20 + 3.

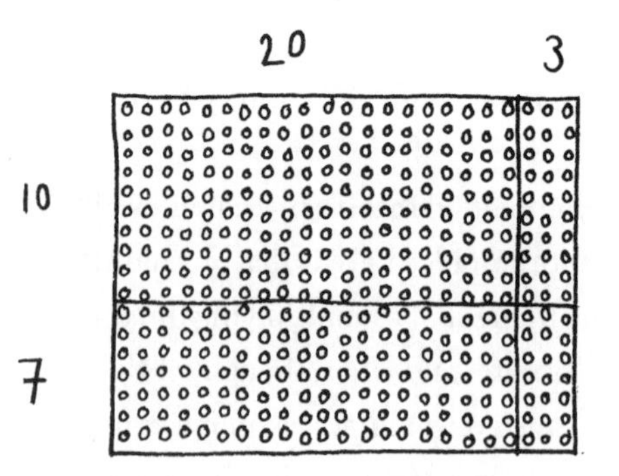

What we have done here is break the large numbers into smaller, easier numbers (known technically as *partitioning*).

Counting the bottles now becomes a matter of working out

how many there are in each block and then adding the four blocks together:

$10 \times 20 = 200$
$10 \times 3 = 30$
$7 \times 20 = 140$
$7 \times 3 = 21$

Total $= 200 + 30 + 140 + 21 = 391$

All we have done here is multiply together each of the numbers in these brackets $(10 + 7)$ by each of the numbers in these brackets $(20 + 3)$.

$$(10 + 7) \times (20 + 3) = 10\times20 + 10\times3 + 7\times20 + 7\times3.$$

Now the connection with algebra becomes clearer, because this way of counting up in blocks will be true for any numbers. Replacing them with letters, a, b, c, d:

$$(a + b) \times (c + d) = a\times c + a\times d + b\times c + b\times d.$$

Or in conventional shorthand, with the multiplication symbols removed: $(a + b)(c + d) = ac + ad + bc + bd$.

If you need convincing, pick any numbers to represent a, b, c and d (1, 2, 3 and 4, say) and put them into the equation to check that both sides come to the same answer.

	a	b
c	ac	bc
d	ad	bd

The Disappearing Multiplication Symbol

As we have already mentioned, when you progress from straightforward arithmetic to algebra, one of the first signs that you've moved on to 'grown-up' maths is that the multiplication symbol disappears. To recap, this is for the simple reason that a multiplication sign looks like an 'x'. '7 multiplied by x' is written simply as '$7x$', and if you want to multiply expressions that involve brackets, such as '5 times $(x + y)$' you write '$5(x + y)$'. When two variables a and b are multiplied together the result is ab. Numbers and variables can be mixed together, but it's normal to put the numbers at the beginning and the variables (usually letters) at the end. For example: $3a$ multiplied by $2b$ written out in full is $3 \times a \times 2 \times b$.

Without the multiplication symbols, and with the numbers and letters grouped together in alphabetical order, it is written as $6ab$. (Since the order doesn't matter when you multiply, it could be written as $6ba$, and in theory even as $ba6$, but it never is.)

Another bit of shorthand that features throughout algebra is when a number is multiplied by itself. 7 multiplied by 7 equals 7 squared, written as 7^2. In the same way, the variable x multiplied by itself is x^2.

From now on in any algebra we write, we will leave out the multiplication symbol.

x times *x* – a common teenage error

Many teenagers really struggle with the idea of x multiplied by itself. A common wrong answer for *x times x* is $2x$, even though they know that 7 times 7 is 49, not 14.

Multiplying (out) the Brackets (or 'Expanding')

Multiplying (out) the brackets simply means getting rid of the brackets just as we did in the example on page 89. Teenagers will often be asked to multiply out the brackets when there are two terms in one bracket and two terms in the other, like this:

$$(a + b)(c + d)$$

Of course a, b, c and d could be any letters that you like, and they can be numbers too. So for example the same rules apply to:

$$(y + 2)(y + 6)$$
and $(3 + a)(b + c)$

When there are two terms inside each set of brackets, multiplying out involves four calculations, just as it did with calculations such as $(10 + 7)(20 + 3)$.

$$(a + b)(c + d) = ac + ad + bc + bd$$

All the combinations that are being multiplied together have been indicated with lines. We've come across many teenagers being encouraged to memorise this using all sorts of devices, including: 'two crossed eyebrows (joining ac and bd) and a smiley face (joining ad and bc)'. But as with other memory aids, the risk of learning this way is that the aid can be misremembered. Surely it is better to understand why it works by just going back to the grid method and, if necessary, replacing the letters with simple, small numbers, to check the answer makes sense.

For example, you can check that:
$(2 + 3)(1 + 5) = (2 \times 1) + (2 \times 5) + (3 \times 1) + (3 \times 5) = 30$.

Incidentally, there is no limit to how many values can be inside a set of brackets. Think about a crate of milk bottles that is 13×24. The bottles can be counted by grouping them as $(10 + 3)(20 + 4)$ or as $(5 + 5 + 3)(20 + 4)$, and indeed a myriad of other ways too. It's then a matter of carefully making sure that you multiply together each of the numbers in one bracket with those in the next (the equivalent of making sure you add up every group of milk bottles in the whole crate).

A word of warning: the more you partition the numbers (i.e. split them up into parts), the easier it is to miss one by mistake when adding them up. Multiplying out $(x + 3)(a + b + c)$ you should end up with six different items (or *terms*) within the expression: $ax + bx + cx + 3a + 3b + 3c$.

TEST YOURSELF

Multiply out the brackets:

a) $7(a + b)$
b) $(x + 3)(x + 4)$
c) $2a(4 + a)$
d) $(a + b)^2$
e) $(x - 1)(2a + b + 3)$

Simplifying

As the name suggests, simplifying is about trying to put algebraic expressions into their most compact, simplest, form.

Why bother? For a start, it's always easier to understand something, and to get the big picture, if you group together all the items of one type. 1 big tin + 3 small tins + 2 small tins + 4 big tins can be expressed much more simply as 5 big tins and 5 small tins. Algebraic simplification is much the same: $5a + 3c + 2a - 4a + 6c$ is easier to grasp when written as $3a + 9c$.

A second more subtle reason for knowing how to simplify algebra is that often different solutions to a problem will give different-looking algebraic answers. Simplifying the answers helps you check if these answers are just different ways of saying the same thing. For example, suppose you are a garden landscaper who specialises in square ponds surrounded by a border of flagstones, like these:

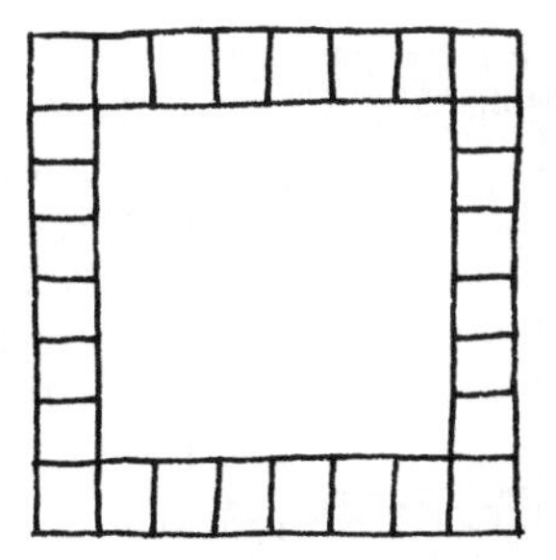

The pond above happens to have sides that are six flagstones long (plus the corner squares), but the pond could be any size 'n'.

There are several ways of thinking about how to calculate the number of flagstones needed. You can imagine laying n flagstones for each side of the pond and then four single ones at each corner. Thought of this way, the total number of flagstones needed is: $4n + 4$.

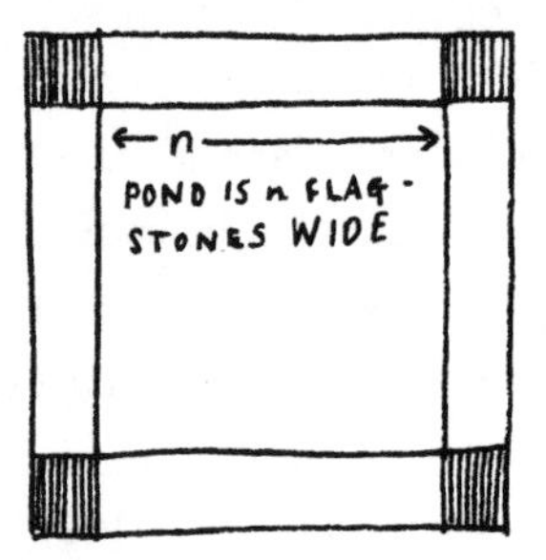

Or you could imagine the sides being made up of four lengths of flagstones each $n + 1$ long. Looked at this way the number of flagstones will be $4(n + 1)$.

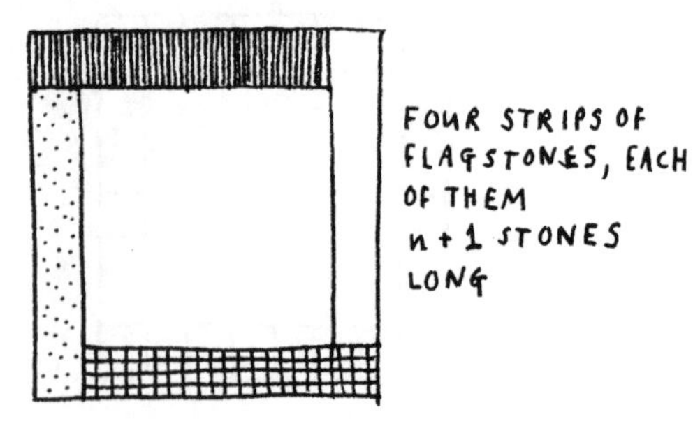

Or you can think of the top and bottom as being $n + 2$ flagstones, while the two sides (shaded) are the length of the pond, n.

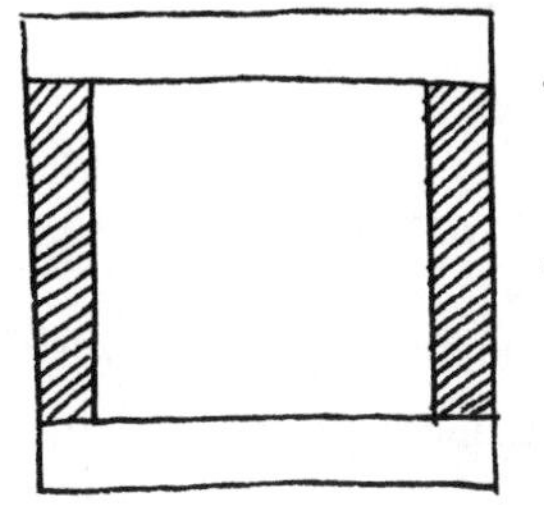

In other words, the total number of flagstones here is

$2(n + 2) + 2n$.

We've come up with three different ways of counting the flagstones:

$4n + 4$;

$4(n + 1)$

and $2(n + 2) + 2n$

They are all describing the same flagstones so they ought to give the same answer. But do they? Multiplying out $4(n + 1)$ does indeed give $4n + 4$, but what about the third rather messy expression?

$2(n + 2) + 2n$

$= 2n + 4 + 2n$

Which simplifies to:

$4n + 4$

It's a matter of taste as to which is 'simpler' $4(n + 1)$ or $4n + 4$. Many people might choose the latter as it doesn't contain brackets (and so is simpler) but mathematicians are likely to prefer the expression $4(n+1)$ because it reduces the answer to a multiplication – usually referred to as *factorising*. More of which in a second . . .

Deliberately making algebra complicated

Understanding the idea of simplifying can be helped by doing the opposite: complicating! On the back of an envelope put a simple algebraic expression. You can almost turn this into a game, 'play' with the algebra to make it look as complicated as possible, while making sure that the actual value is the same. For example:

$4n + 4$

Subtract $3n$ then add it back in: $4n - 3n + 3n + 3 + 1$.

Reorder and tidy up: $n + 1 + 3n + 3$

Multiply everything by 6 and divide by 6: $6[(n + 1) + 3(n+1)] \div 6$

For the really adventurous, multiply out the bracket and split the fraction: $6(n+1) \div 6 + 18(n+1) \div 6$

If you choose a value for n, for example $n=1$, you should find that you get the same answer (8 in this case) whichever of the expressions above that you use.

Factorising

Your teenager first met the idea of factors in primary school (and very likely forgot them again soon afterwards). The factors of a whole number are those whole numbers that divide into it exactly. For example, the factors of 12 are 1, 2, 3, 4, 6 and 12, while the factors of 18 are 1, 2, 3, 6, 9 and 18.

Two or more factors can be multiplied together (their *product*) to make the original number. For example, among a multitude of different ways, you can form 18 from factors like this:

3×6

or 2×9

or $1 \times 3 \times 6$

or $2 \times 3 \times 3$

This last expression: $2 \times 3 \times 3$ is particularly important – it is the *prime factorisation* of 18. In other words, 18 has been expressed as the *product* of only prime numbers. It's important because these factors cannot be broken down into any smaller factors. (As a reminder, a prime number is a number that is only divisible by itself and by 1; in other words, it is a number with exactly *two* factors. For example, 7 only divides by 1 and 7, while 23 only divides by 1 and 23. The number 1 is not prime because it only has one factor – itself!)

Factorising in algebra is the same idea as factorising numbers. It means writing an algebraic expression as the product of two or more 'basic' units. A longer way of saying 'factorise' would be: 'work out which simpler expressions would be multiplied together to make this one'.

For example, take the expression:

$4n + 8$

What, when multiplied together, makes $4n + 8$? To find out, look for what you can *divide* it by. Everything in the expression can be divided by 4, so when factorised* it becomes: $4(n+2)$.

In fact, you'll notice that factorising is effectively the opposite of multiplying out the brackets. Factorising means putting the brackets back in!

Factorising 'quadratic' equations

Factorising becomes more challenging when it involves squares or higher powers. For example, factorise: $3x^2 + 12x$. In other words: 'What simpler expressions when multiplied together make $3x^2 + 12x$?'

The expression $3x^2 + 12x$ is known as a *quadratic* because it contains a variable (x) that has been squared (quad meaning squared, as in *quad*-rangle).

In this example everything divides by 3, so 3 is a factor. We can rewrite the expression as: $3(x^2 + 4x)$. Can we find another factor, in other words is there anything else that divides into $(x^2 + 4x)$? Yes, everything divides by x.

So we can write out the expression as: $3x(x + 4)$.

That's it, we've reduced this to the most basic factors we can have. The original expression $3x^2 + 12x$ has three factors, 3 multiplied by x multiplied by $(x + 4)$.

If only all quadratics were so easy to factorise. Sadly, they aren't.

What are the factors of: $x^2 + 5x + 6$?

It is certainly not obvious what simpler expressions can be multiplied together to make this one.

* It's the convention to factorise as far as possible. $4n+8$ could be 'factorised' as $2(2n + 4)$, but the $(2n+ 4)$ can also be divided by 2, so $4(n+ 2)$ is the fully factorised version.

We will tell you the answer, and then explain how to find it.

The factors of $x^2 + 5x + 6$ are $(x + 2)$ and $(x + 3)$. You can check this by multiplying out the brackets: $(x + 2)(x + 3) = x^2 + 2x + 3x + 6 = x^2 + 5x + 6$.

In fact, any expression of the form 'x^2 + a multiple of x + a number' (like the one we just saw) will always be the product of exactly two simpler expressions, '$x + a$' and '$x + b$', where a and b are numbers. The challenge is to work out the values of the numbers a and b.

Factorising a quadratic using the grid method

Remember the grid method (page 87)? You can use it as a method for factorising a quadratic. The expression $x^2 + 5x + 6$ has two factors of the form 'x + a' and 'x + b' – we won't attempt to prove why, so you have to take this on trust. How do you find out a and b? You can set out the multiplication as a grid:

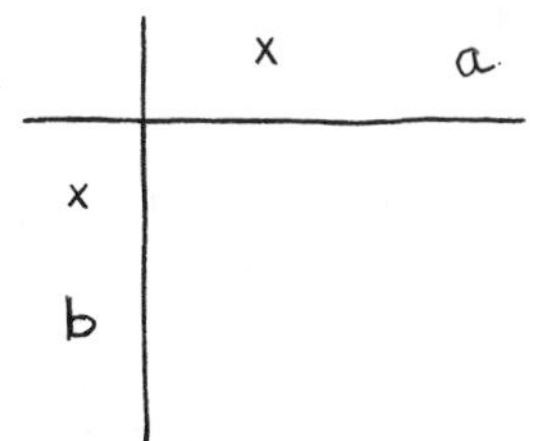

and then multiply out the grid in the same way as you would numbers, like this:

	x	a
x	x^2	ax
b	bx	ab

Unlike normal grid multiplication, in this case we already know what the answer needs to be. We have been told that the whole grid adds up to $x^2 + 5x + 6$. Now it's time for some detective work.

The only place in the grid where we have a number on its own (as opposed to some multiple of x) is ab. This means that a multiplied by b must equal 6 (so a and b could be 1 and 6, or 2 and 3, for example)

Next, look for multiples of x. We have $ax + bx$, and this must equal $5x$, so $a + b = 5$.

So, we know $ab = 6$ and $a + b = 5$. What are the possibilities?

How about $a = 1$ and $b = 6$?: $ab = 6$ but unfortunately $a + b = 7$ – so this doesn't work.

$a = 2$ and $b = 3$?
$ab = 6$ and $a + b = 5$ – yes, it works!

So the solution must be: $a = 2$ and $b = 3$ (or vice versa), in other words:

$$x^2 + 5x + 6 = (x + 2)(x + 3)$$

We've just factorised a quadratic – praise the Lord!

Your teenager might be taught a particular method for factorising quadratics (there is in fact a standard formula), but it does no harm to be aware of the 'reverse grid method' approach we just used because you can go back to it when the formula has been long forgotten.

Why do you need to factorise equations?

This still leaves the uncomfortable question of *why* is there any need to factorise equations in the first place? Like many of these algebra tools, factorising comes into its own in more advanced maths. If your teenager goes on to do A level, they will be grateful for a good grounding in factorising when they meet calculus and even probability theory, both of which can involve a lot of algebra. Factorising is often a helpful way of simplifying an equation, and can make it easier to plot a graph (see quadratic equations on page 125). And factorising often provides a shortcut to solving a problem. It might be used to work out the maximum height that a rocket reaches before falling to earth or the best route for a telephone cable if you want to use as little wire as possible. If your teenager never goes on to those more advanced uses, take comfort from the fact that some teenagers really enjoy factorising just for the sake of doing it, just like a Sudoku.

Learning how to factorise takes a lot of practice. Your teenager's textbook or worksheets will have numerous examples. Our aim here has simply been to show you what it's about, and give at least a hint as to why it's needed.

How to square numbers in your head – using factorising

Factorising can also lead to general discoveries about numbers that were not otherwise obvious. For example, imagine a square number such as 5^2 (= 25) or 8^2 (= 64). You can represent any such square number as x^2. And if you want a number that's one smaller than a square, write it as $x^2 - 1$.

How do you factorise $x^2 - 1$?

It's not obvious, so we'll tell you that: $x^2 - 1 = (x + 1)(x - 1)$

What this tells us is that if you pick any number x, such as 5, or 99, or 2507, you can be sure that the square of your number x^2, for example 5^2, will be one more than $(x + 1)(x - 1)$, in this case 6 multiplied by 4. Sure enough, $5^2 = 25$, and $4 \times 6 = 24$, which is one less. Try it for another number, 11 for example. $11^2 = 121$, and $10 \times 12 = 120$, one less. It always works.

So in order to square any number x, you can if you like multiply together $(x - 1)$ and $(x + 1)$ and then add one. It might be hard to square 99 in your head, but it's easy to multiply $98 \times 100 = 9800$, and then add one. In other words $99^2 = 9801$. (You might want to read that again, slowly.)

This arithmetical shortcut can be done for all numbers. What is 999^2? Easy: it's the same as 998 times 1000 add one = 998001. So surprising as it might seem, you can now amaze your friends at parties by squaring numbers such as 9999 or 100001 in your head. $(x + 1)(x - 1)$ does have its uses.

TEST YOURSELF

1) Using the technique described opposite, work out in your head the following:

a) 19^2
b) 101^2

2) Factorise the following expressions

a) $3a + 3b$
b) $2 + 8q^2$
c) $x^2 + 5x + 4$
d) $x^2 + 5x - 6$

Rearranging Equations

Algebra problems start to get really tough when you have to rearrange them in order to come up with a solution. Rearrangement can crop up in several different types of maths problem, for example:

a) Find x if $x + 7 = 4x - 2$

Or b) Rearrange this equation to make x the subject:

$$y = \frac{(2x + 10)}{7} - x$$

(We'll solve these in a moment.)

What can prove to be so challenging about problems like these is not that the maths is particularly difficult, but that these problems often require several steps, each requiring a solid understanding of at least one mathematical principle.

Think of 'equations' as balancing scales

Always think of an equation as being like a balance. What's on the left is equal to what is on the right. The equals sign acts as the pivot.

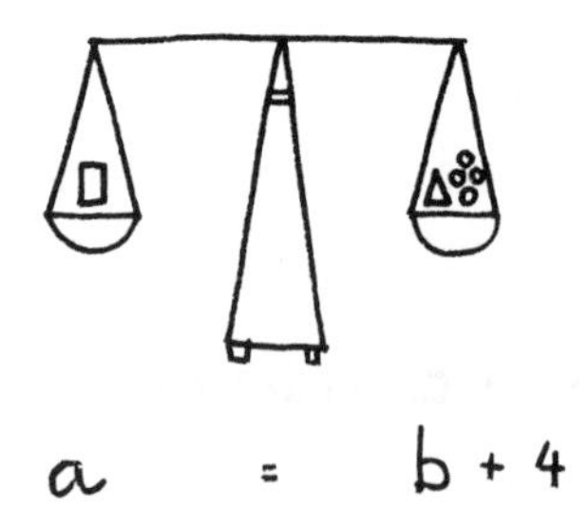

$$a = b + 4$$

The idea of balancing scales is a powerful one for understanding equations because of a simple principle that always applies: if two sides balance, then if you do something to the left-hand side *and do exactly the same thing* to the right-hand side, the two sides will still balance.

Whenever your teenager is struggling to grasp a problem involving rearranging equations, go back to the balancing principle. A physical prop can help, so if you have a set of old-fashioned balancing kitchen scales, use them.

Balancing works for all the mathematical operations: if you add to one side, add the same amount to the other side. Look at the example above. You can see that a equals ('balances with') $b + 4$. You can now do anything to both sides, for example add 2, and it will now be true that $a + 2 = b + 6$. It will also be true that a + Fred = $b + 4$ + Fred . . . or anything else you care to add or subtract. If you double one side, double the other and all will still balance; if you divide one side by three, divide the other side by three. It even works with more complicated operations such as squaring: if you square one side then square the other and it will still balance. So if $a = b + 4$, then it's also true that $a^2 = (b + 4)^2$.

(For the balancing analogy to be strictly accurate, the items being balanced need to be put on pans that are both the same distance from the centre, or *fulcrum*. Otherwise, using the lever principle you could get any two unequal weights to balance just by moving the heavier one closer to the centre!)

So how about those examples we met earlier?

a) Find x if $x + 7 = 4x - 2$

'Find x' means we want to know what 'x equals'. So we need an equation which has '$x =$' on one side, and an expression that doesn't have any 'x' on the other.

There are a number of ways to do it, but whichever approach we use, it will involve always doing the same to both sides of the 'balance'. For example:

$$x + 7 = 4x - 2$$

1) Add 2 to both sides: $\quad +2 \qquad +2$

$$x + 9 = 4x$$

2) Remove one x from both sides $\quad -x \qquad -x$

$$9 = 3x$$

3) Divide both sides by 3 $\quad \div 3 \qquad \div 3$

$$3 = x$$

. . . and we have the answer.

Incidentally, instead of 'Find x', mathematicians sometimes write 'Solve for x'. It means the same thing.

Here is the second example we set earlier:

b) Rearrange this equation to make x the subject:

$$y = \frac{(2x + 10)}{7} - x$$

'Make x the subject' again means rearrange this equation so that it becomes '$x =$' on one side, with NO xs on the other side. Here's one way to do it:

1) Multiply both sides by 7: $\quad 7y = (2x + 10) - 7x$
   (A typical teenage error is forgetting to multiply the '$-x$' by 7. Remember, you have to multiply EVERYTHING on both sides by the same amount.)
2) Simplify: $\quad 7y = 2x + 10 - 7x = 10 - 5x$
3) Add $5x$ to both sides: $\quad 7y + 5x = 10$
4) Subtract $7y$ from both sides: $\quad 5x = 10 - 7y$
5) Divide both sides by 5: $\quad x = \frac{10 - 7y}{5}$

And there's the answer!

Many teenagers are unhappy with an answer like this, reasoning that they haven't really 'solved' what value x has to be, that the answer $(10 - 7y) \div 5$ is still 'uncertain'. But in fact it is a perfectly good answer: what it is saying is that x is a variable that can take a range of answers, depending on y. For example, if $y = 0$, then $x = 2$, if $y = 1$, $x = 0.6$ and so on.

By the way, your teenager might be familiar with different language when manipulating algebra, for example they might have phrases like: 'take it to the other side and change the sign'. So, for example when you add $5x$ to both sides of $7y = 10 - 5x$ it becomes $7y + 5x = 10$. It certainly looks like $5x$ has been 'taken to the other side and its sign changed' but that's really the end result of balancing the equation by doing the same to each side. Both forms of language are correct, but for understanding what is going on, it is more useful to use the 'balancing' language.

Simultaneous Equations

If teenagers find solving one equation difficult, how about trying to solve two at the same time? Simultaneous equations are regarded as sufficiently daunting that they don't appear in the more basic GCSE curriculum. Yet as with so much in algebra, the idea behind them is very straightforward. An equation gives you some information about a variable, such as the price or the area of something. Sometimes, however, you might be given *two* pieces of information, two clues if you like, and you need to piece those together in order to work out an answer. (It's usually true that if you are given two clues it means you can work out two numbers, with three clues you can work out three numbers, and so on.)

In fact an example of this appeared early in the book, with the infamous 'Bow and Arrow' question on page 5. As a reminder: *The bow and arrow cost £11 together, and the bow costs £10 more than the arrow. How much does the bow cost?*

Once they've overcome the shock that the answer is *not* £1, some teenagers do attempt to solve it, and most often they will use trial and error. They spot that £1 is too much for the arrow, and £0 is too little, and guess (correctly) that the answer will be in the middle, 50 pence.

Trial and error is fine for a simple problem like this, but it can become tediously slow for more complicated problems. Simultaneous equations offer a quicker method of finding a solution.

In the case of the Bow and Arrow puzzle, there were two equations hidden within the question:

$$B(\text{ow price}) + A(\text{rrow price}) = £11$$

$$B \qquad - \qquad A \qquad = £10$$

There are two unknowns (the bow and arrow prices) and there are two equations that apply at the same time, hence they are 'simultaneous'.

If we know what the bow costs then we can work out what the arrow costs and vice versa, but at the moment we know neither. The idea behind simultaneous equations is to eliminate one of the unknowns using the sort of manipulations we encountered earlier on. In this case there is an easy way to eliminate the cost of the arrow. In one equation there is a '+ arrow price' and in the other there is a '− arrow price' so these two can cancel each other by adding together the two equations.

$B + A = £11$

$+$

$B - A = £10$

To get:

$2B + A - A = £21$

$2B = £21$

And hence the cost of one bow is £10.50.

Now we know the cost of the bow, we can put that value back into either of the equations to work out the cost of the arrow (50p).

Simultaneous equations get much harder than this. In order to eliminate one variable you might have to multiply one or more of the equations, subtract negative numbers and the rest. And there can be more than two equations. The key thing is that as long as you have as many equations as there are unknowns, and as long as each equation gives you different information (is 'independent') then you can solve equations of any size.*

Your teenager might not be expected to do much on simultaneous equations up to GCSE, but here's a harder example as a reminder of what is involved:

Find x if $3x + 4y = 25$ (1)
and $2x + y = 10$ (2)

* For example, $3x + 4y = 25$ and $12x + 16y = 100$ look like a pair of simultaneous equations, but they are not – the second equation is simply the first one multiplied by four, so it doesn't give any new information.

Step 1: We need to get rid of y since we've been asked to find x. To do that, we need to have same number of ys in each equation so that we can subtract one set from the other.

Step 2: Multiply equation (2) by 4: $8x + 4y = 40$ (3)

Step 3: Subtract (1) from (3): $(8x + 4y) - (3x + 4y) = 40 - 25$

Step 4: Simplify $5x = 15$

Step 5: Hence $x = 3$

(And if $x = 3$, then using either (1) or (2) you can work out that $y = 4$.)

TEST YOURSELF

a) A hat and coat cost £300. The coat costs four times as much as the hat. What does the hat cost?
b) Find x if $3x - y = 20$ and $x - y = 6$

Using Algebra

Many teenagers are only comfortable using algebra when it relates to something in the 'real world'. Fortunately, there are plenty of formulae in maths that have practical applications, from the area of a circle ($A = \pi r^2$) to the conversion from miles to kilometres ($K = 1.6M$).

Algebra is also essential for creating 'mathematical models' of the world. Engineers, economists and weather forecasters use maths to simulate the real world all the time, and in doing so they will be creating formulae.

There are more examples of real-world algebraic formulae in the next chapter.

TEST YOURSELF

Two mobile phone companies are offering different charging schemes. When making a call, both of them charge a fixed amount per month plus a charge per minute of phone use. The first company, MegaCall, charges £20 per month, plus 10p per minute. The second company, NanoPhone, charges £10 per month plus 15p per minute.

a) If you use the phone for 100 minutes per month, which company offers the better deal?
b) Write an equation for the total cost per month for the two phone companies.
c) How many minutes would you need to use the phone per month for it to cost the same whichever phone company you used?

Summary

TERMINOLOGY	MEANING	EXAMPLE
Expand (or *'multiply out the brackets'*)	Get rid of the brackets (by multiplying together).	$(x + 2)(y + 3) =$ $xy + 2y + 3x + 6$
Factorise	The reverse of 'expanding', put the brackets back in.	$x^2 + 3x + 2 =$ $(x + 1)(x + 2)$
Simplify	Group similar terms together and (usually) make an expression as short as possible.	$3x + 5 + 4x + 2$ simplifies to $7(x + 1)$
Solve for x	Turn an equation in which terms with 'x' appear in more than one place into an equation of the form '$x =$' with NO xs on the other side.	$x + 3 = 2x - 4 + y$ 'solved' becomes: $x = 7 - y$
'Make x *the subject'* Or *'Write* x *as a function of* y*'*	Rearrange an equation so that it is in the form '$x =$ [something to do with y]'.	$y = 7x - 3$ rearranges to: $x = (y + 3) \div 7$
Substitute for x	Work out the value of a function when x is a particular value.	$x^2 - 4x + 7$ when $x = 2$ has the value 3

ALGEBRA
If you do only four things . . .

- Use the example of milk bottle crates (page 87) to remind your teenager how multiplication works. You can use grid diagrams to represent the multiplication of algebraic expressions.
- When in doubt, replace letters with (small) numbers to test that an equation works. And encourage your teenager to do this after they have simplified or rearranged expressions.
- Think of equations as a balance. Manipulating algebra is all about 'doing the same thing to both sides of an equation'. If you do the same thing to both sides, the equation will still balance.
- Explain that manipulating algebra is an exercise in logic and paying attention to detail. These are useful skills regardless of whether or not your teenager goes on to higher maths (and if they do, they'll find these basic skills are essential tools).

FUNCTIONS AND GRAPHS

In all the discussion of algebra in the two previous chapters, something has been missing: the big picture. In the case of algebra, the 'picture' is a graph, a visual representation that can bring abstract ideas to life.

Graphs can be informative (for example, they can show when something has an upward trend, or when it has reached a peak) and with their sinuous curves they can even be pretty, but another important use is as a visual aid for problem-solving. The solution to a tough algebra problem can become instantly obvious when it is presented as a graph.

For example, without giving you any clues at all, which point represents the 'solution' to the question behind these graphs?

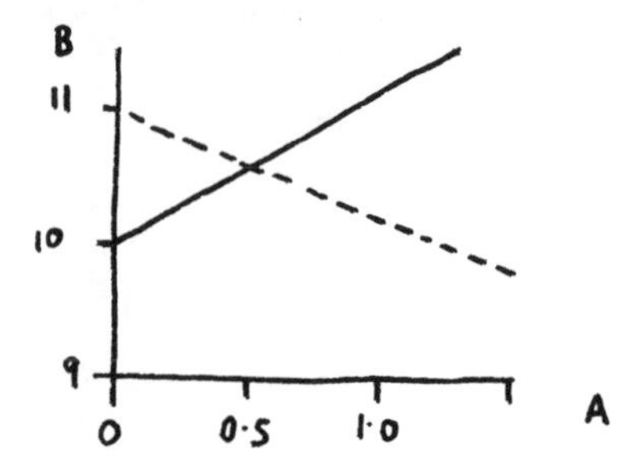

No doubt your eye is drawn to the point where the two lines cross – it looks like a case of 'X marks the spot'. The co-ordinates of the crossing point are $A = 0.5$ and $B = 10.5$. And this is indeed the solution – but to what?

In fact this is a graphical solution to the Bow and Arrow puzzle (page 5). A represents the price of the arrow and B represents the price of the bow and the point where the lines cross represents the solution where the bow (B) is £10.50, the arrow (A) is 50p. As we'll explain shortly, what we've done here is solve a *simultaneous equation* (see page 107) but we've done it visually rather than by manipulating algebra. And you would be right in thinking that many teenagers might find this visual way of solving equations a lot easier.

The Bow and Arrow graph explained

The dotted line represents the statement 'the bow and arrow cost £11 in total' (bow price + arrow price = 11). Look at any point on the dotted line and read off the values on the vertical and horizontal axes A and B. At every point on that line A and B add up to 11 (for example, when B = 10, A = 1 or when B = 11, A = 0).

The solid line represents 'The bow costs £10 more than the arrow' (or bow price = arrow price + 10). At every point on the line, the value of B is 10 greater than the value of A. (For example, when B = 11, A = 1.)

The two lines meet at the point where both statements are true, and that is the solution.

X-Y Functions

How do you start to build algebraic graphs to solve problems? One way your teenager might be taught is through the idea of 'functions'. Although the two axes on a graph can represent anything (bow and arrow price, temperature, distance travelled), they usually carry the rather less inspiring labels of the x-axis (which is horizontal) and the y-axis (vertical).

It is also normal to regard x as the 'input' and y as the 'output', the output being the quantity that we want to work out; y is then described as being 'a function of x'.

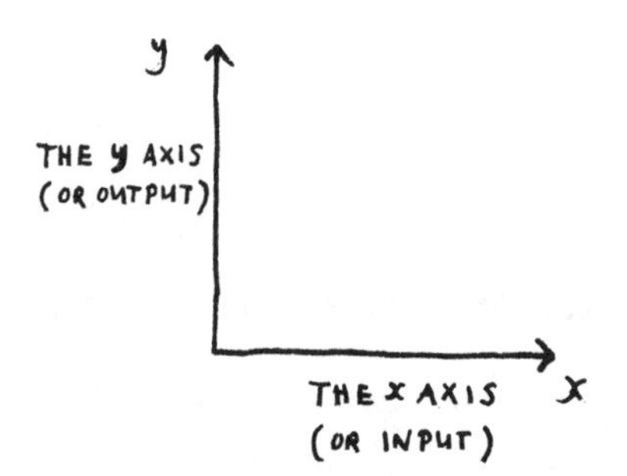

One popular way to think about functions is as a factory where the input goes in at one end, has a 'function' done to it, leading to the output at the far end.

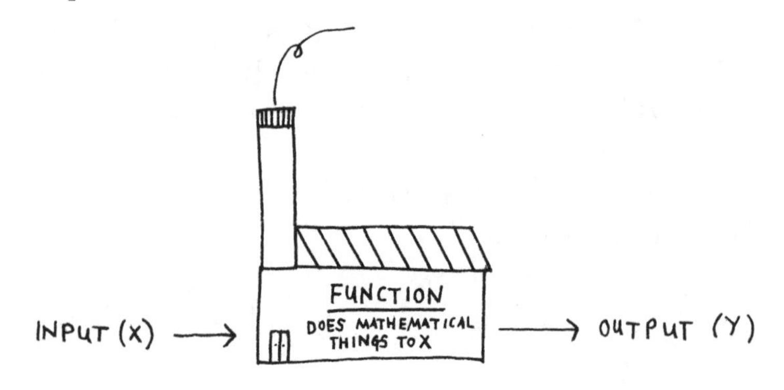

This 'function' can be as simple as just adding a number. In the case of the bow and arrow puzzle, if we treat the arrow price as the 'input', then add £10, we get the bow price as the output.

Arrow price -> add £10 -> bow price.

A slightly more complicated example is the mobile phone charges from the Test Yourself question on page 111. In this case you use the amount of time spent on the phone (the input) to work out how much your bill is going to be (the output).

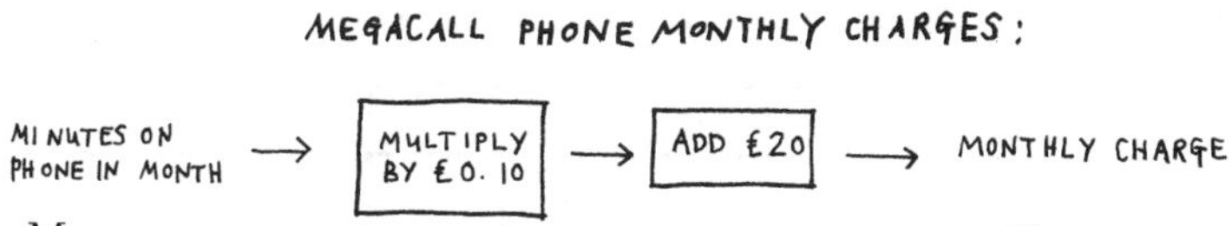

MEGACALL PHONE MONTHLY CHARGES:

MINUTES ON PHONE IN MONTH → MULTIPLY BY £0.10 → ADD £20 → MONTHLY CHARGE

M C

In other words: $C = 0.1M + 20$.

Using this idea of a flowchart also means you can reverse the order to come up with a formula for M in terms of C, so C becomes the input and M is now the output. All you need to do is reverse the steps and reverse the mathematical operation at each point.

Start at the right and move to the left:

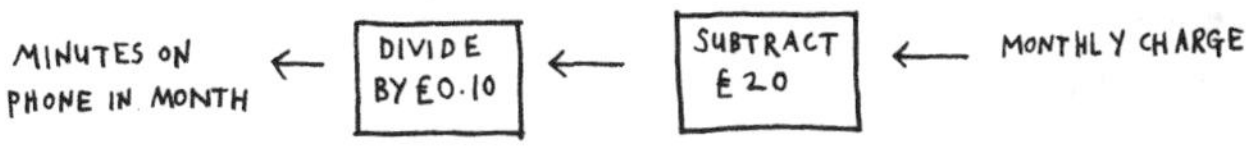

MINUTES ON PHONE IN MONTH ← DIVIDE BY £0.10 ← SUBTRACT £20 ← MONTHLY CHARGE

The reverse function is $M = (C - 20) \div 0.1$ and you can use this to work out how many minutes you have used the phone for when presented with the monthly charge. (This procedure works for basic functions, but be careful, it starts to get messy with things like quadratic equations – more of which later.)

Plotting functions

You can take any function like the one for mobile phone charges and simply plot it as a graph. As a stepping stone, however, teenagers will be introduced to the idea of working out several values of the input and output, and entering them into a table – either manually or, increasingly often, using a spreadsheet on a computer.

The output (monthly charge) can be tabulated for different values of M:

Minutes (M)	× £0.1	+ £20 = C
0	0	£20
100	£10	£30
200	£20	£40
300	£30	£50

These values can be plotted on a graph:

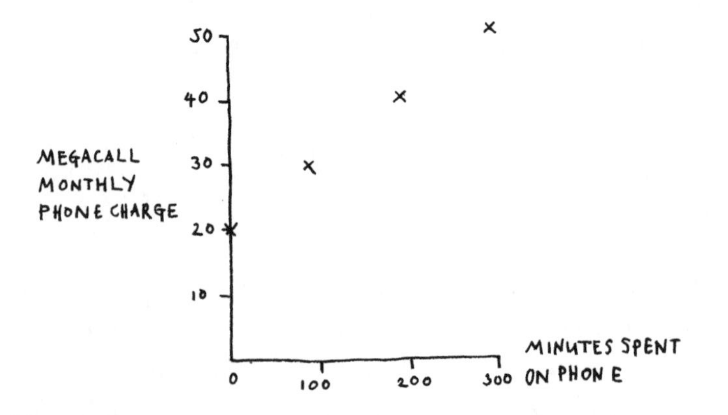

Now it's just a case of 'joining the dots' with (in this case) a nice straight line. You can immediately read off the cost of any number of minutes by looking up and across at that value. The cost of

150 minutes? That will be £35. (Note, however, that this is only possible because time is 'continuous'. If the x-axis were, say, number of phone calls made per month, then strictly speaking a line graph doesn't make sense, as it would mean you could read off the cost of 'half a phone call' – an impossible concept!)

Just as with the Bow and Arrow problem, graphs can be used to solve the mobile phone problem on page 111, where you were challenged to find what level of phone usage would give you the same bill whether you use MegaCall or NanoPhone. The first company, MegaCall, charges £20 per month, plus 10p per minute. The second company, NanoPhone, charges £10 per month plus 15p per minute. The formulae for the two companies are therefore:

MegaCall charge $= 0.10T + £20$ (where T is the number of minutes)

NanoPhone charge $= 0.15T + £10$

The values for M and N can be worked out for different values of T, and plotted straight onto a graph:

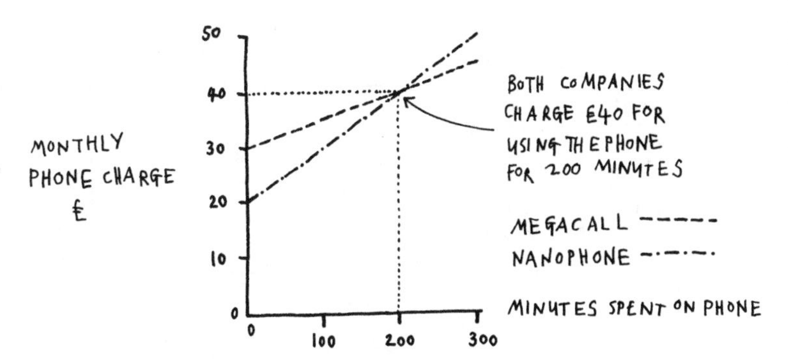

The two lines cross at 200 minutes, where the cost for each service is £40 – the solution to the problem. The graphs also tell you something about what happens before and after the crossing point. Below 200 minutes, NanoPhone charges are always lower than MegaCall. Above 200 minutes, the position is reversed. A teenager who is on the phone all evening should be encouraged to use MegaCall . . .

Linear Graphs, and $y = mx + c$

The examples we've seen of graphs so far are all straight lines, and the two variables (call them x and y, or the input and output), have what is called a *linear relationship.* This pattern crops up all over the place. Here are some other everyday examples of straight line graphs:

FORMULA	APPEARANCE OF GRAPH
ROASTING TIME FOR TURKEY = 30 MINS + 40 MINS PER KG*	
TAXI FARE = DISTANCE TRAVELLED × MILEAGE RATE + FLAG FARE†	
TEMPERATURE IN FAHRENHEIT = 1·8 × CENTIGRADE + 32	
WEIGHT OF CRANBERRIES = 2 × WEIGHT OF SUGAR (FOR CRANBERRY SAUCE)	
INCOME TAX = BASIC RATE × (INCOME – ANNUAL ALLOWANCE)	
AVAILABLE CREDIT = BANK LIMIT – AMOUNT SPENT	

* Depending on the oven temperature etc. The authors take no responsibility for food-poisoning or other disappointments should you decide to use this rule of thumb.

† The flag fare is what they charge for simply climbing into a taxi.

These graphs have two important features in common: they slope by a fixed amount (that's what straight lines do, after all), and there is some point at which they meet the vertical axis, known as the 'intercept'. For *four* of the graphs (turkey, taxi fares, temperature and bank credit) the intercept point is above zero. For cranberry sauce, the graph passes exactly through zero on both axes, a point that is known as the *origin* (zero cranberries means zero sugar is needed, of course). In the case of income tax, the line stops before it reaches the vertical axis (you can't have 'negative income tax'), but if you do extend the line, it passes through the vertical axis at a negative value.

What does the intercept represent? It's the starting value when the input is zero: for example, when Celsius is zero degrees, the intercept of the Fahrenheit axis is 32°. But what does the slope represent? It is the rate at which the output increases compared to the input. The time to roast a turkey increases by 40 minutes *per extra kilogram.* The taxi fares increase at a rate of £2.50 *per mile.* In mathematical-speak, these graphs are all of the form:

$$y = mx + c$$

where m is the slope (or *gradient*) of the graph, and c is the point at which it crosses the vertical axis. The reason why c is used is that it stands for 'constant', because it is some fixed number. The reason for calling the slope m, on the other hand, has been lost in the mists of time.

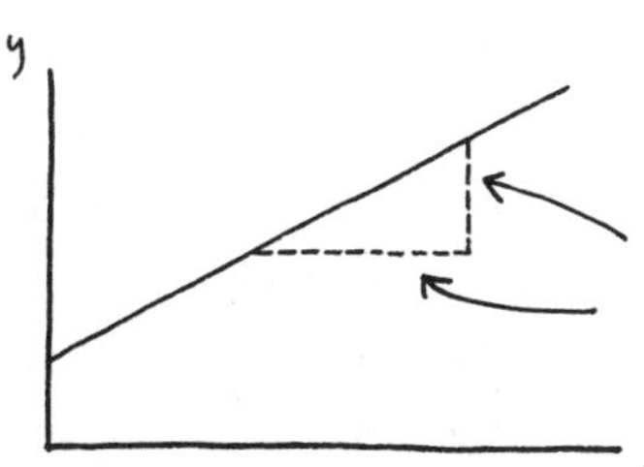

The slope m and the intercept c can be any combination of positive, negative or zero. As it happens, in all the examples above *except for the bank credit*, there is a positive gradient because the two variables are directly connected – as one increases so does the other. An inverse relationship is one where one variable decreases as the other increases; for example, as the amount you spend on a credit card increases, the available credit decreases. This is why the credit graph has a negative gradient that slopes down from left to right.

The position of the intercept varies in our examples on page 120. It is positive for turkey, taxis, temperature and credit, zero for cranberry sauce, and negative for income tax. Here are some typical values for the graphs:

	Slope '*m*'	Y-intercept '*c*'
Turkey roast	40 mins per kg	30 mins
Taxi fares	2.50 (£ per mile)	£2.00
Celsius to Fahrenheit	1.8°F per °C	32 °F
Cranberry sauce	2	0
Income tax	0.2 (20%)	−£2,000
Available credit	−1	£3,000

Now straight lines are not difficult to understand, so why do straight-line graphs cause grief for so many teenagers? The main difficulty is that a graph can be looked at and considered in two ways: as a representation of a relationship between two real-world quantities (as we have been using it); and also as a mathematical object in its own right. And it's when graphs become abstract mathematical shapes with 'gradients' and 'intercepts' that teenagers' eyes begin to glaze over. This is why it can be useful to have real-world examples like the ones we've given to fall back on when explaining the concepts.

TEST YOURSELF

Champion sprinters like Usain Bolt run 100 metres in about 10 seconds.

a) Sketch the graph of the distance run by Usain Bolt against time.
b) What is the slope of the graph?

Global warming gone mad

In 2002, when global warming was beginning to really hit the news, Reuters put out this news item: 'The Antarctic Peninsula has warmed by 36° Fahrenheit over the past half-century, far faster than elsewhere on the ice-bound continent or the rest of the world.'

John Shonder, an engineer in Tennessee, found it hard to believe there could have been such a huge rise in temperature, so he looked up the original announcement of this story by the British Antarctic Survey. What they had said was: 'During the last fifty years the Antarctic Peninsula has warmed by 2.5° C.'

The Reuters journalist had used the formula for converting Celsius to Fahrenheit: $F = 1{\cdot}8\,C + 32$. This formula works if you want to convert a temperature. For example, plug 0° C (freezing point) into the formula and you get 32° F, which is correct. But if you want to work out the *change* of temperature, it helps to look at the graph. The slope of the graph in this case is $9 \div 5$ or 1.8 degrees Fahrenheit per Celsius. If the temperature increases by 2.5°C, the Fahrenheit equivalent increases by $2.5 \times 1.8 = 4.5$°F, which is rather less alarming than the 36° reported.

Common Teenage Errors

1) **Misinterpreting the meaning of a 'steep' gradient.** In the graphs above, the slope of the graph for taxi charges is much steeper than the slope for roasting a turkey. Does this mean that taxi charges are more 'extreme' in some way? Maybe that explains why taxi fares are so high? Well . . . no. We have deliberately left the scales off all of those graphs, which means it isn't possible to compare the relative extremeness of taxi fares, roast turkeys, income tax or anything else with those graphs. The only meaningful comparison of graphs is when *both graphs* are drawn using the same axes with *the same scales* on each axis – a misunderstanding that newspapers often take advantage of, by presenting graphs side by side with different scales.
2) **Thinking 'proportional' and 'linear' mean the same thing.** For making cranberry sauce, if you double the number of cranberries you double the amount of sugar you need. Those two ingredients are in direct proportion to each other and their relationship is also linear. But this is

not the case for any of the other graphs. Doubling the size of turkey doesn't double the time it needs to go in the oven, for example. Thinking that linear and proportional are always the same is a very common error, but the only time when the two are equal is in those graphs where there is no constant added so that the line passes through zero. (The story of the 'global warming' scare on page 123 was caused by a journalist making this error.)

3) **Assuming everything 'steady' is linear.** Teenagers tend to think 'when in doubt, assume it's linear'. It's understandable, after all the shortest distance between two points is a straight line, but it's often wrong – as we'll see later.

Parabolas and Quadratic Equations

Not everything in life has a simple, linear relationship.

Mankind has known for thousands of years that if you drop something off a cliff it speeds up, but it wasn't until the early 1600s that Galileo started plotting accurate graphs of what was going on. What Galileo discovered was that the distance that an object had fallen increased not in proportion to the time an object had been falling, but in proportion to the time *squared*.

The number pattern of squares is something that children first encounter in primary school:

1×1 = 1	2×2 = 4	3×3 = 9	4×4 = 16	5×5 = 25

What they don't do until they are teenagers is plot this on a graph. The shape is a pleasing curve.

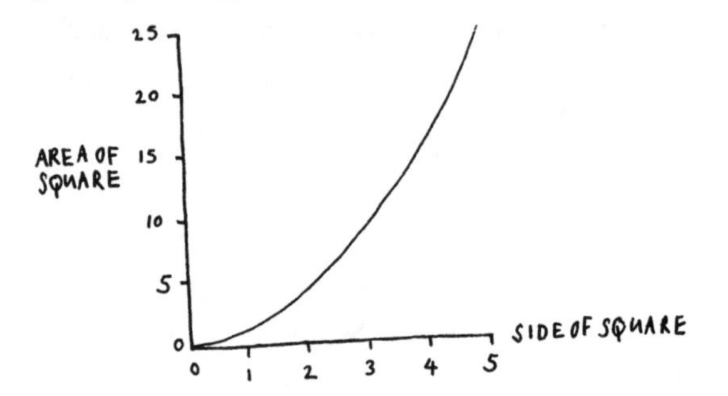

Its name is a *parabola*. It is connected to the word parable (as in the stories told by Jesus) since both words come from the Greek word 'parabole', which means 'running alongside' or 'a comparison'. Quite what 'running alongside' has to do with the mathematical shape nobody is quite sure.

If you turn a parabola upside down, you get the path of a ball lobbed in the air. Needless to say, scientists in Galileo's time were surprised (and delighted) that something from nature such as a falling object should have a connection with squares, and the discovery ultimately led to Isaac Newton formulating the universal laws of gravity.

The equation for the curve is $y = x^2$. The big difference between this and the earlier functions is that it involves *x squared*. As we saw in the previous chapter, any equation that involves *x squared* is known as a *quadratic* equation,* the 'quad' part being because of this link to squares.

We've actually only shown half of the graph here because we've ignored the negative numbers, but there's no reason why we

* A quadratic equation always involves x^2 and it can also contain a multiple of x and/or a number. If it contains any other power, such as x^4, it is not a quadratic. The letter x can be replaced by other letters, of course.

can't continue the axes into those regions. What happens to the parabola with negative numbers? Since a negative number multiplied by a negative number becomes a positive (see page 203), $-1 \times -1 = 1$, and $-2 \times -2 = 4$, and so on, in fact we produce a mirror image of the first curve. The base of a full parabola looks like this:

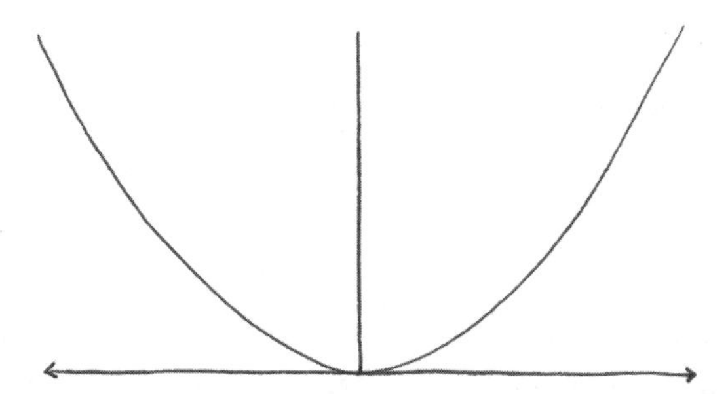

It forms a cup. For the graph of $y = x^2$ the parabola's base sits at the point 0,0.

Adding quadratics to straight lines

We're about to discuss something that is on the edge of what teenagers might be expected to handle, but these are important ideas that can help to join up the topic of algebra and graphs.

If you are the type of person who likes to push the boundaries a little, you might be curious to know what happens if you mix equations involving x^2 with linear equations that just involve x (such as the one for roasting turkeys, which was $y = 40x + 30$).

You might think that $y = x^2 + 40x + 30$, which we'll call lobbed-ball-plus-roast-turkey, would produce a graph that is some horrible contortion of a straight line with curvy bits, something of a turkey, in fact. But this is not the case. The shape of $y = x^2 + 40x + 30$ is in fact a parabola, with an identical shape to the one on page 127 – the only difference is that it's been shifted 20 units to the left and 370 units down (370 seems a surprisingly huge number to come out of such simple equations).

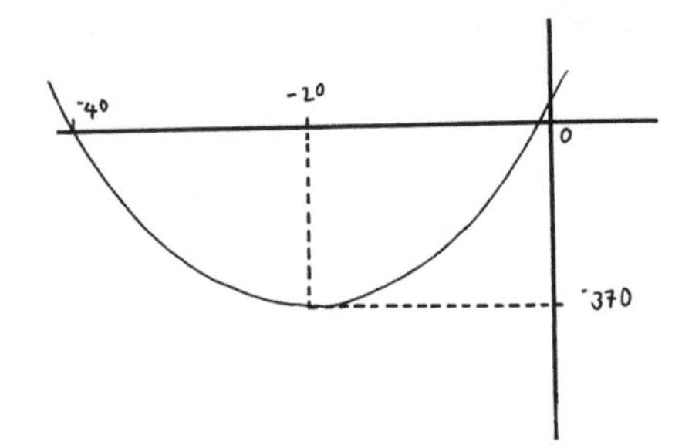

In fact, *any* equation of the form $y = ax^2 + bx + c$, where a, b and c can be any numbers you choose, is a quadratic equation. If you plot it, the result will always be a parabola that is the same as the one above only shifted up, down, left or right. If the value of 'a' is bigger than 1 then the parabola will be stretched, and if a is negative the parabola will be upside down. But it's still a smooth, symmetrical parabola. *Always*.

This surprising and rather beautiful fact is rarely brought to the attention of teenagers. Instead, they spend most of their time trying to figure out how and where to plot these graphs.

How are teenagers expected to find out where a parabola crosses the horizontal axis? The slow way to find out is to produce a table and plot the results manually (with a spreadsheet this is considerably quicker, but it's still a bit laborious).

The faster way, with practice, is to *solve the quadratic equation.* In this case, 'solve' means 'find the point at which y is zero in the equation you've been given' (such as $y = x^2 + 40x + 30$). This is the same as asking, '*When does the graph cut the horizontal axis*'? (Because the horizontal axis is the line where $y = 0$.)

Sometimes a parabola cuts the axis in two places, meaning there are two different values of x for which $y = 0$ (two solutions). Sometimes it touches it only once (meaning there is only one solution). And sometimes it doesn't touch the horizontal axis at all (which means there is no solution).

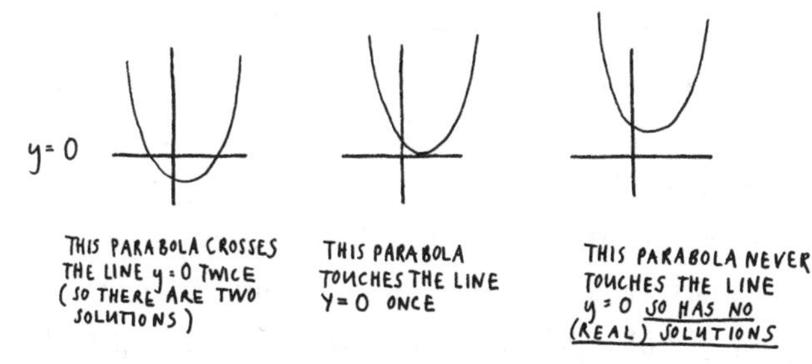

Many GCSE students never even meet quadratic equations, and the formal method for solving quadratics is beyond what we can cover in this book. (Any textbook and countless websites will explain the technique.) But it is still interesting to know what is going on, especially since these equations have a very physical connection to the world around them.

Six places to find a parabola

1) Find a circular cone (a card rolled into a dunce's cap cone is better than a crumbling ice-cream cone) and cut a slice parallel to the side of the cone. The shape of the cut edge is a parabola.

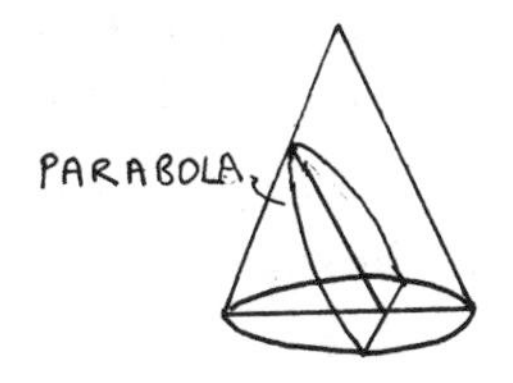

2) Find your nearest suspension bridge. The curve formed by the suspension cable is (very close to) a parabola.
3) Get a copy of the Highway Code and look up the table that tells you the stopping distances for different speeds. Plot on a graph the distance versus speed and the result is a parabola. (Most people might expect it to be a straight line. What it demonstrates is that when driving, a little more speed means you need to allow a much bigger gap between you and the car in front.)
4) The cross section of a satellite dish is a parabola (albeit a very flat one). The reason is down to an important mathematical property of parabolas, which is that lines approaching a parabola in parallel will reflect off and meet at a focal point. This means that the satellite dish can pick up weak signals and concentrate them into a much stronger one, improving the television reception.

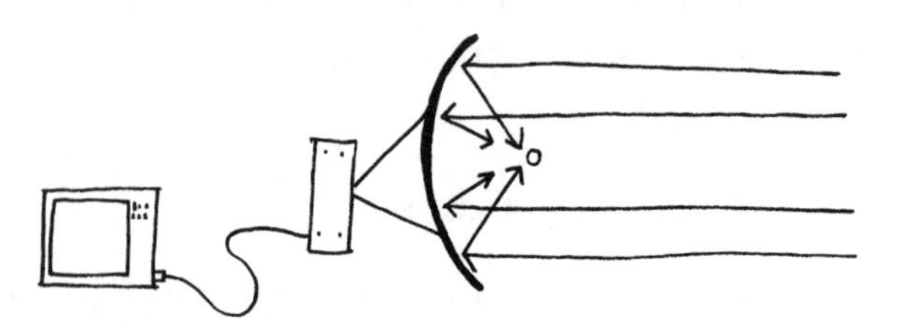

5) The path of the water spraying from a hose forms a parabola as does an Angry Bird being catapulted at pigs.
6) To make your own parabola, put a large sheet of paper on a board and lean the board at an angle. Now dip a marble in black ink, and roll it up the slope. The marble will leave a paint trail which will be the shape of a parabola.

Exponential Curves

There's an important type of curve that looks a little like a parabola, but in fact is fundamentally different. It's the exponential curve, and it applies to anything in life whose rate of growth is proportional to how large it is already. Money earning compound interest is an example: if the interest rate is 10% per year (nice thought!) then a deposit of £100 will grow to £110 the first year, then £121, then £133.10, growing by an increasing amount each year. Animal populations also tend to grow exponentially, at least when the population is small: the more animals there are, the more babies there will be.

An extreme exponential curve might be one where the number doubles every unit of time (every hour, say). This can be true of certain types of cell or bacteria:

Hours (x)	0	1	2	3	4	5	6	7	8
Number of cells	1	2	4	8	16	32	64	128	256

Written as a formula, the number of cells in this case is $N = 2^x$ (we say more about powers on page 206). This looks a little like the basic quadratic formula we met on page 127, $\mathrm{N} = x^2$. All that has happened is that the x and the 2 have been swapped over. But the impact of this little change is huge. Here's what the two graphs look like:

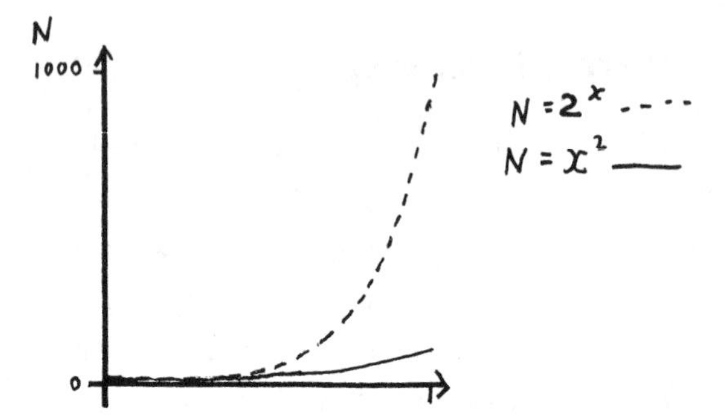

Remember the monthly allowance challenge on page 14 where a penny grew to several million pounds in one month? That growth followed the same exponential curve. Exponential functions are astronomically different from parabolas. Just about the only thing that they have in common is that in their usual form they both curve upwards.

Sketching Graphs

The algebra behind graphs can get extremely difficult. Before long, you can be looking at maths that is well beyond GCSE and often into degree level. And yet a basic understanding of the shape of all sorts of graphs is an important skill for all teenagers, as they will encounter many of these patterns in adult life. Indeed, we'd argue that the ability to 'sketch' a graph is ultimately as important a skill as being able to plot it precisely.

Here is a simple example of a sketching exercise: *A man leaves home, walks to the supermarket, buys various items, enough to fill two heavy carrier bags, then staggers home again. Sketch a graph of his distance from home against time.*

There are several basic graph principles embedded in this story. The man starts at a distance zero and his distance from home then increases steadily (we may as well assume he takes a direct route to the supermarket as we aren't told any different). Then he effectively stands still for a while as he shops so his distance from home doesn't change. Then he staggers home – at a slower rate than when he went. The graph should look something like this:

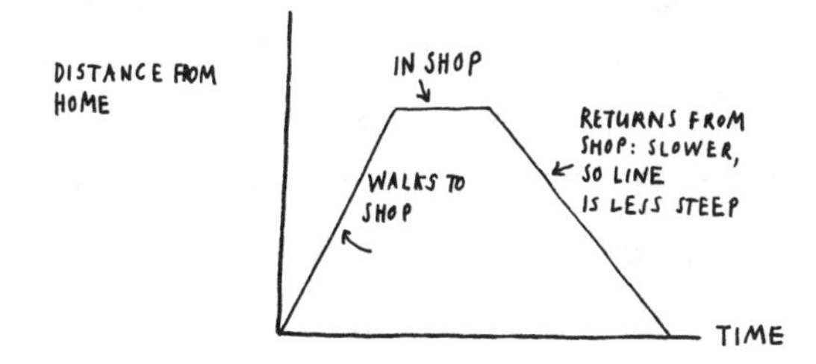

A typical teenage error will be to treat a graph too literally. For the return journey they will often have the line turning back on itself – forgetting that this would mean the man would now be travelling back in time. Some even think that this graph represents going up a hill and down the other side.

Here's an example that on the face of it looks similar to the idea of walking to the shop: *A tap is used to fill a pig's trough. Looked at from the side, the trough is the shape of a V. The trough starts empty and after a minute it is full. Sketch a graph of the height of the water in the trough during the minute that it is filling, by joining the points that mark empty and full.*

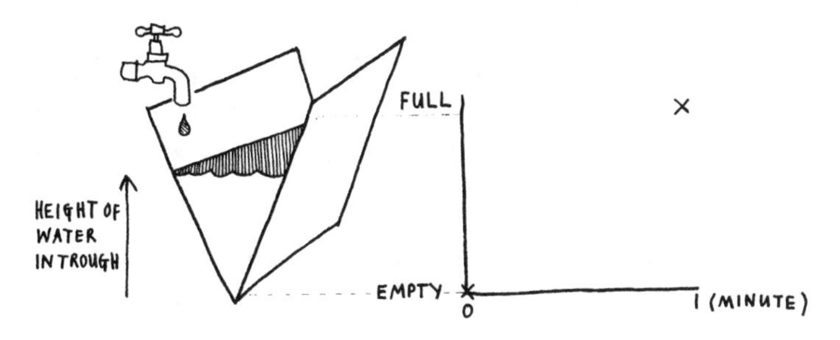

What do teenagers do? They see the straight sides of the trough and, naturally, almost all of them assume that the graph is a straight line, too.

A moment's reflection reveals why this can't be correct. The diagram above shows the trough filled to half of its height and it's clear that the *volume of trough* that still has to be filled is much larger *than the small V that has been filled so far*, and so it's going to take a lot longer to fill the top 'half' of the trough than the bottom half. In other words, the trough fills to half its height quite quickly, but the rate at which the water level rises slows down. That means the graph is not a straight line but a curve. It will look something like this:

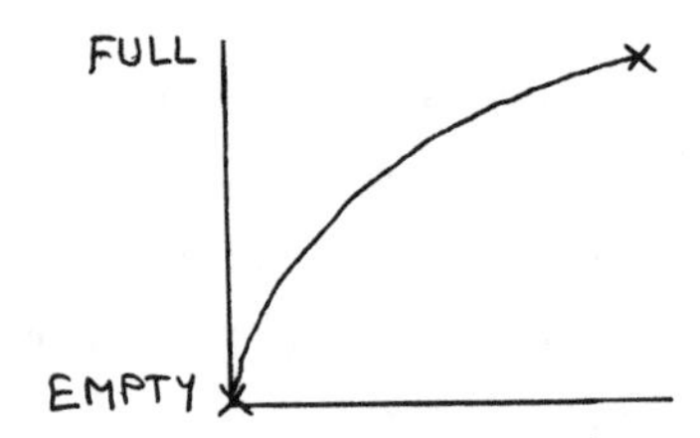

In fact, this is half of a parabola, the graph we met on page 126.* But the detail is not important here, what matters is the general idea of the shape.

Let's take another 'real world' example. *Imagine you are travelling in a Ferris wheel, and that it takes the wheel four minutes to do a complete turn (one cycle). What does the graph of your height above the ground look like?*

Some points are easy to plot: at zero and four minutes, you are at ground level. After two minutes (halfway through the ride) you are at the highest point. And after one minute, a quarter of the

* Though instead of being of the form $y = x^2$, it is actually $x = y^2$ which is the normal parabola *after a quarter turn*.

journey, you are halfway up (a quarter turn). We've plotted these points:

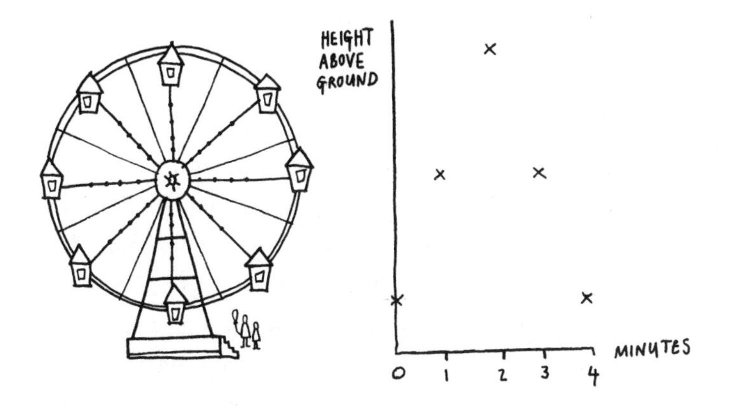

The points plotted so far form a triangle and it's tempting to join them up to form that shape (many teenagers do). But again, a little reflection shows that the shape of the graph will be more complicated. At the start of the ride, which direction are you going? Sideways – gaining no height at all! For the first few seconds you remain almost level, then you start to rise more steeply, and finally at the top of the ride you are momentarily travelling sideways again. This means that at zero and two minutes the graph has almost no slope, while at one minute the rate of ascent is at a peak so the graph will be at its steepest. In fact, the graph looks like this:

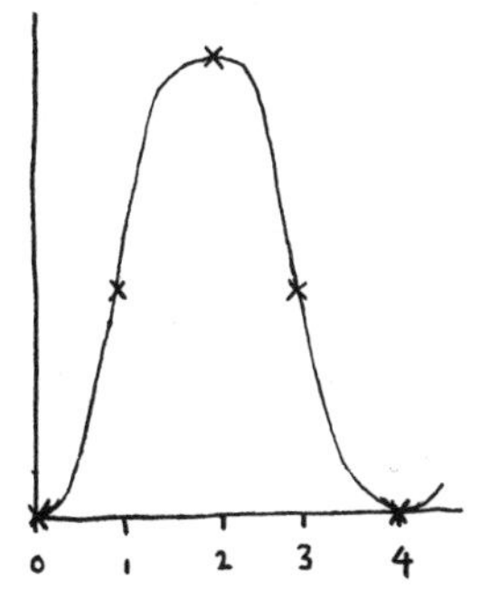

This wave shape is known as a 'sine wave', and if you stay on the Ferris wheel beyond one cycle, your height maps out a repeating sine wave that could go on for ever.

The sines are everywhere

You might recognise the word 'sine' as something to do with triangles (see page 171) and there is indeed a connection. If you pick one of the spokes of a Ferris wheel, it forms the long side of a triangle (the hypotenuse), and if you plot the height of the vertical side of that triangle as you move the spoke around the circle the graph is a sine wave. Teenagers are taught plenty about circles and triangles up to GCSE, but few are made aware of how closely the two are connected to each other.

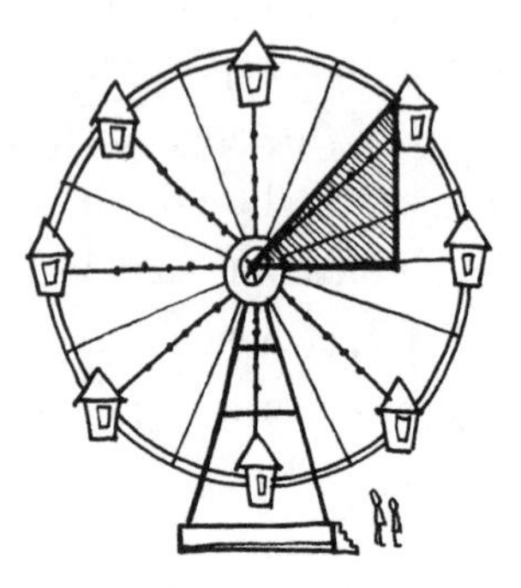

Creating graphs to fit the real world

The graphs in this chapter started from 'neat' mathematical formulae, and it was then a matter of plugging in the numbers to see the shape of the graph. Life isn't always so tidy, however. If your teenager goes on to do maths to A level and beyond, they will find themselves looking at data from a real-life situation (such as the annual sales of a company) and then trying to find a mathematical function whose graph best fits the data. The most common shapes of graph that are used to model the real world in this way are straight lines, parabolas and exponential curves. And having found a graph that appears to fit the data, that graph can then be used to predict what might happen. If you hear any news story forecasting future quantities of anything, from people to products, those numbers almost certainly came from people with a solid grounding in the algebra of graphs.

FUNCTIONS AND GRAPHS
If you do only three things . . .

- Encourage your teenager to learn how to do rough sketches of graphs. These can be quite whimsical: your emotions over the course of a day, the speed of a car going round a race track, and so on. The important reason for doing this is to see that graphs don't always resemble the 'look' of real-life experiences. (By far the most common type of graph is something being plotted against time.)
- Look out for graphs in the news, on TV or in magazines. Use them as a talking point, and see what conclusions you can draw from them.
- If your teenager has an interest that involves data that could be plotted, use that as an excuse to generate graphs on a spreadsheet at home. It could be as simple as a graph of their savings and spending each month, or their height. A teenager will take far more interest in a graph if it contains information of direct interest to them.

GEOMETRY

A line wants to know about the future of his love life, so goes to visit an astrologer. The astrologer looks into her crystal ball. 'Ah,' she says, 'I see a parallel line coming from the mist . . . you two have so much in common.' Then her face turns sad. 'Unfortunately, the two of you will never meet.'*

* Joke given to us by a 13-year-old, who wished to remain anonymous.

What *IS* Geometry?

If a well-educated dad* from ancient Greece were to be transported to twenty-first-century Britain, the only part of a secondary-school maths exam that he would recognise would be the geometry questions. In his era, geometry and maths were really the same thing.

In fact, thanks to a man called Euclid, who in around 350 BC produced an astonishing and definitive text called *The Elements* (running to thirteen books, each packed with dozens of theorems), geometry continued to dominate maths education until Victorian times. If your great-grandparents went to a grammar school, they probably studied Euclid till he was coming out of their ears and they would look at the modern geometry syllabus and scoff that it is barely scratching the surface. On the other hand, most would also accept that today's teenagers have a rather more enjoyable time with maths than they did. And they would also be stumped by some of the other topics, such as probability, that they didn't have time for in their curriculum.

In simple terms, geometry is the study of shapes, in particular their angles, lengths and areas. These things can, of course, be measured with a ruler and a protractor, which is the practical side of geometry beloved of surveyors, builders and the people who make curtains. But what really gets a geometrician excited is the stuff that doesn't require any sort of measuring device, where you can work out lengths, angles and stuff *just by using reasoning*, and where the only items you need to draw a perfect shape are a straight edge and a pair of compasses. This (ancient) form of

* Hardly any mums studied maths in those days.

geometry is more like puzzle-solving, and in fact it has a lot in common with tackling a Sudoku: you solve the big problem by finding an easy point of entry – 'what do I know, and what can I easily figure out?' – and then making a series of much simpler little deductions, each of which leads on to the next.

Geometry in its widest sense is deeply linked to many other areas of maths, including ratios, fractions and algebra. We've therefore put three chapters under this general heading of Geometry:

- Properties of shapes, and how one shape compares to another.
- Triangles.
- Moving and changing shapes.

PROPERTIES OF SHAPES

There are two reasons why the study of shapes is included in the school mathematics curriculum:

1) **To learn how to construct two-dimensional (2D) and three-dimensional (3D) geometrical shapes.** Since shapes are a fundamental part of engineering, graphics and art, this skill clearly has some possible practical applications, though 'mathematical' construction is more about the thinking than the doing, as we'll explain.
2) **To help teenagers to develop their mathematical reasoning and the idea of proof.** This has little to do with shapes at all. There are plenty of other subjects, not just in maths, that can train the mind in reasoning. But some educators regard the elegance, conciseness and purity of the reasoning used in geometry as the best 'brain-training' there is. For this reason, if you believe in brain-training on the Nintendo DS, maybe you should give geometry a second chance.

Let's describe these two different uses a little further.

Constructing geometrical shapes

How do you make an equilateral triangle? With your computer it's trivial, just click on a shape, but even with pencil and paper it's easy enough: get a ruler, draw a line 10cm long (say), then get a protractor, measure an angle of 60° at each end of the line, join them up and you have it.

If you're good at using drawing instruments then this equilateral triangle that you have produced should be accurate. It might even be incredibly accurate. But it is not mathematically *precise* – accuracy and precision are different things. However careful you are in creating a triangle, there is bound to be some inaccuracy, it all depends on how closely you follow the markings on your ruler and protractor . . . and, for that matter, it depends on how accurately the manufacturers printed the markings onto the ruler in the first place. Even if you're using a computer to create the equilateral triangle, can you be sure that it is accurate down to the last pixel?

Now imagine a world where nobody had ever invented centimetres or degrees or any other measurement scale, and all you had was a compass and a straight edge. Amazingly, it is still possible to create many regular shapes, and these shapes aren't accurate, they are precise.

Mathematical precision is a Big Idea in maths, and one that many teenagers can find hard to grasp, especially since, when it comes to a mathematician's precision of an equilateral triangle, the only precision is in the imagination – it can't ever be done in practice. The construction of an equilateral triangle was the very first piece of geometry tackled by Euclid in Book 1 of his *Elements*. Here's how he did it:

1. Imagine a line AB: A ——— B

2. Draw two circles with a radius of AB. One of the circles is centred on A, the other on B. (To make this real, you can stick a compass in A and draw the circle, then do the same at B.) You end up with this:

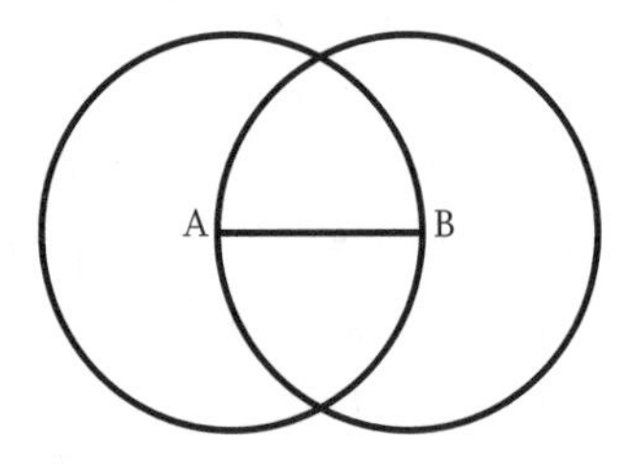

3. Call the higher point at which the two circles meet C. Join A to C and join B to C with straight lines:

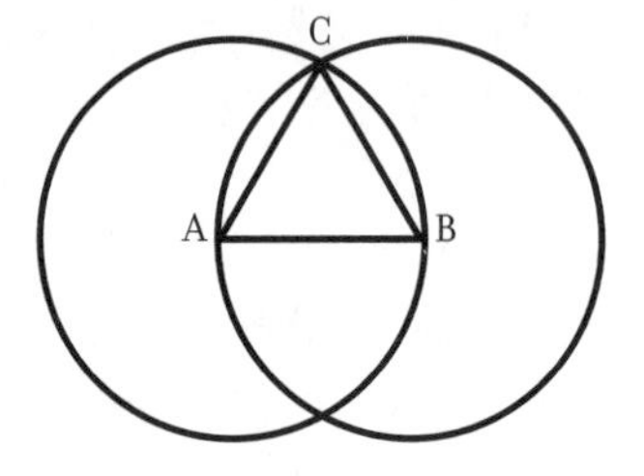

4. You have created a triangle. The base of the triangle is AB, which is the radius of both circles that we just created. What about AC? Since A is the centre of the circle and C is on the perimeter, the line AC is *also* the radius of that circle. So AB = AC. Meanwhile B is the centre of the second circle, and C lies on that circle, so BC is also the radius, and AB = BC. In other words, the three lines are the same length so this must be an equilateral triangle. And it will ALWAYS be an equilateral triangle, whatever the length of AB.

Each of the steps here was obvious, but the grand result, that you can construct an equilateral triangle using a line and two circles, was not. It's almost enough to make you want to shout eureka (though the Greek who famously uttered that word was not Euclid but Archimedes, who was also a geometry whizz). This construction of an equilateral triangle crops up frequently in geometry problems. It's also the building block for creating a classic tiling pattern of hexagons. To create hexagons, simply add another circle at point C, then continue to add circles centred on every point where any two circles meet. The hexagonal pattern emerges as if by magic, each hexagon being made up of six equilateral triangles:

What we are doing here is creating precise shapes without the need for any measuring device. The nice thing about precise shapes (as opposed to accurate ones – remember the subtle difference) is that you can describe precise shapes without having to draw them. For example, here is a description of a precise square: it has four straight sides of equal length and four equal angles. No square that you ever construct will be as accurate as that description.

Maybe this is all getting a bit philosophical, but it's important to know that in a way geometry is a philosophical subject just as much as it is a practical one. Some teenagers like the philosophical side of it. Though to be honest, most don't (in fact, most have probably never even given it a second thought).

Geometrical reasoning and proof

This leads us to the second reason for studying shape: it's a way of learning about proof. The word 'proof' is a word that is bandied around a lot in courtrooms and newspaper stories, and is usually used to mean 'evidence' (remember various claims of 'proof' of weapons of mass destruction prior to the Iraq war). This is very different from mathematicians' use of 'proof', which means 100 per cent categorical certainty, with no exceptions. Proof allows you to take an observation and test out whether what you have seen is always true, or is just true in certain situations.

Euclid's *Elements* was really a huge collection of theorems (a theorem is a mathematical statement that has been proved), with each proof building rigorously on the previous one.

Proofs are a difficult concept for most teenagers, partly because they think of them more along the lines of the casual everyday use, as in 'I've tested an example, it worked, so that's proved it.' The maths educator John Mason came up with a tongue-in-cheek way of explaining the three stages of proof for a teenager:

Stage 1 proof: 'I've convinced myself.'
Stage 2 proof: 'I've convinced my friend.'
Stage 3 proof: 'I've convinced my enemy.'

Only when a proof has got to Stage 3 can you be confident that it really is a proof . . . and even then, remember that your enemies can be fooled sometimes.

Angles, Cangles, Fangles and Zangles

Children first encounter angles in primary school, but at that stage it's mainly discovering the different types of angle (acute, right, obtuse and reflex – see the Glossary) and how to measure them with a protractor.

In secondary school, teenagers begin to learn about how the angles in different shapes are related to each other. This is where proofs and theorems begin to appear. And at this point it can all *seem* to get incredibly complicated, especially when pupils are expected to learn theorems. For some teenagers, geometry can become little more than a ritual of memorising a set of rules without necessarily having any understanding of how those rules came about.

We encountered one teenage girl who had been learning a fundamental set of basic theorems. These were linked to the angles between the intersecting lines in the diagram below. We've labelled the various angles with letters a to h:

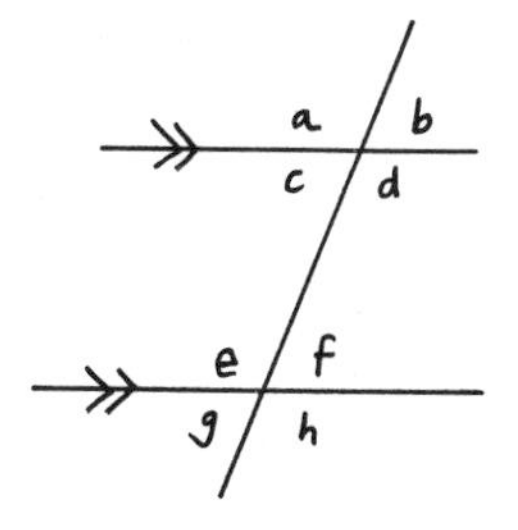

The two horizontal lines are parallel (that's what the arrow markings, known as chevrons, indicate) and because of those parallel lines, it turns out that the angles are all related to each other. In fact, angles a, d, e and h will always be identical, as will b, c, f and g. You might have suspected as much just by looking at them.

The girl we talked to had learned a set of rules to remember how these angles related to each other, using the rather catchy names of Zangles, Cangles and Fangles (quite a lot of schools use these words, by the way). The trouble was, she was getting very confused between her Zangles and her Fangles, and who can blame her because as an exercise in memorising it does seem very complicated. Here's what they are:

First, there are the angles *c* and *f* that form the inside of a Z. These are known as the Z-angles, or Zangles (more formally known as the *alternate angles*).

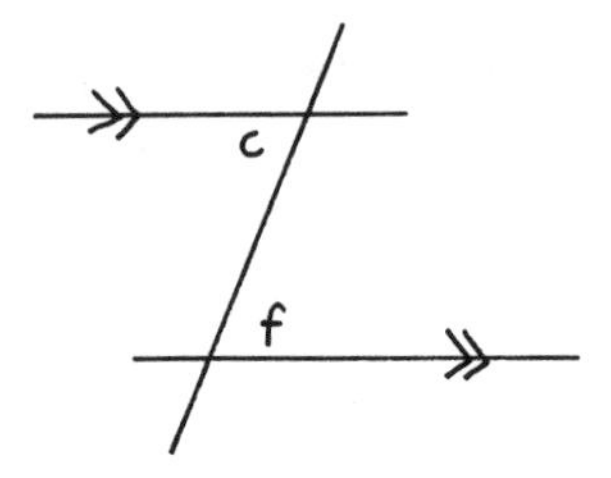

As long as the two lines being connected are parallel, then the two **Zangles are always equal.**

Next come the angles *d* and *h*, that form the joints within a sloped letter F, and are therefore called the Fangles (or more formally, the *corresponding angles*).

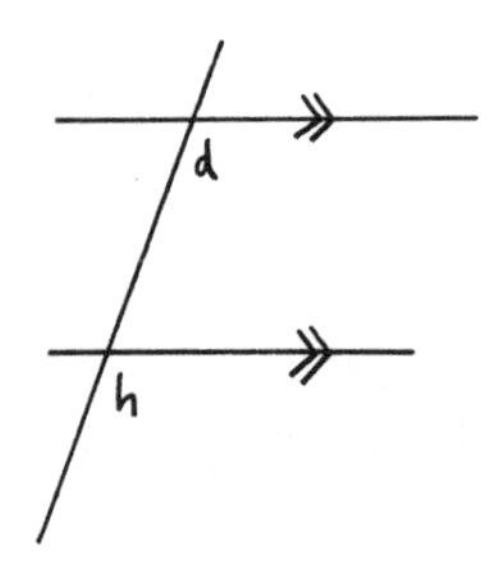

Like the Zangles, the two **Fangles are always equal.**

Finally you have angles d and f that form the inside of a sloped letter C. Not surprisingly, they're called the Cangles, or, less predictably, the *co-interior angles*.

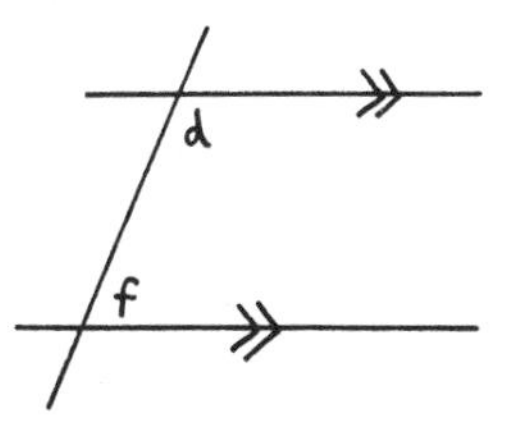

The **Cangles always add up to 180 degrees.** (A pair of angles that add to 180 degrees are known as *supplementary angles.*)

Now memorise that lot, and we'll test you later. OK, maybe not. Millions of pupils survived in previous generations without ever needing to know the *names* of these angles, they simply learned the simple, and after a while intuitively obvious, fact that d and f add to 180° – and the even more obvious fact that d and h must always be the same. The names are only useful for understanding what somebody else means when they use them – otherwise they are frankly little more than annoying and unnecessary jargon that will only confuse the typical teenager.* A much better way to learn the connection between all these angles is to understand the simple reason why these rules work, and then to do lots of examples that involve them until the rules become second nature.

* Also, your teenager may be obliged to use this terminology to get full marks in exams.

How to prove the rules for Zangles, Fangles and Cangles

We'll take you through the simple steps, using the same reasoning that Euclid did. (The only difference is that Euclid didn't use degrees, he just talked about circles and semicircles.)

First, imagine a straight line PQ meeting another line RS at an angle. If you like, think of these as two roads meeting at a T-junction:

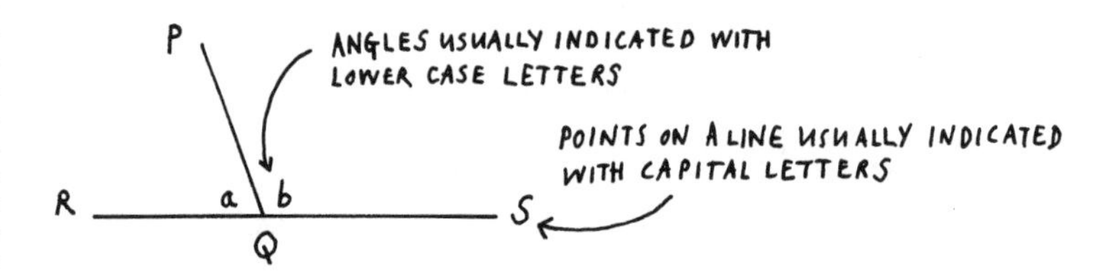

We said PQ joins 'at an angle', but there are of course two angles, one acute (a) and the other obtuse (b). (Notice we've used the convention of labelling angles with lower-case letters and the points on a line with capitals.)

Together the two angles a and b form a straight line, half of a full turn, which is 180°. We'll not attempt to prove this, we'll just state it as a fact that should be obvious (it's what Euclid would have called one of his *axioms*).

Of course a and b still add to 180° even if the line PQ crosses over RS to form a crossroads. If we do extend the line PQ, we create two more angles, c and d.

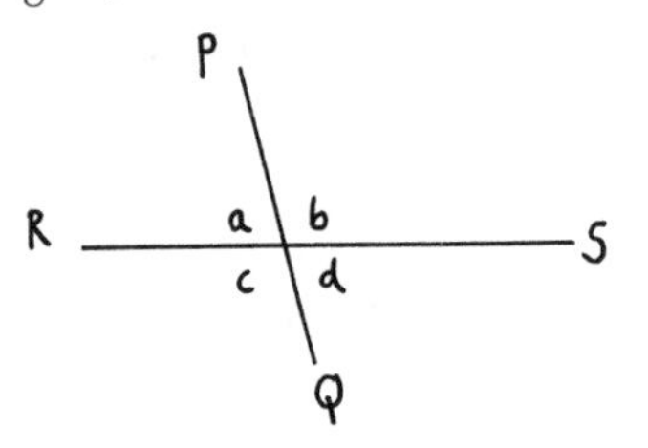

And here is the crucial part: since PQ is a straight line, then the angles a and c must *also* add to 180°.

We already know that $a + b = 180°$ and now we know that $a + c = 180°$. This means that b *and* c *must be equal to each other.*

Using the same argument, since $a + b = 180$ and $b + d = 180$, then a *and* d *must also always be equal*.

In other words:

OPPOSITE ANGLES WHEN STRAIGHT LINES CROSS ARE ALWAYS EQUAL

The reason we put that simple theorem in a box is that it is extremely important, and we'll refer back to it later.

Having proved that opposite angles are always equal, if you look back at the Cangles, Fangles and Zangles, you can see just by inspection why all the rules are true. That's not quite a rigorous proof, but it will do for our purposes.

Make your own Zangles and Cangles

You can generate instant examples of Zangles and Cangles using a pencil and a small envelope (or some other object that has parallel sides). Lay the pencil at an angle across the two parallel edges of the envelope. As you move the position of the pencil, which angles are equal to each other? Which always add to 180°?

PENCIL AND ENVELOPE: WHICH ANGLES ARE THE SAME?

Right Angles

Of all the angles that crop up in geometry, one of the most important is the right angle, which is 90°. Right angles have a vital role to play in the physical world, since they represent the meeting of the vertical with the horizontal, and any builder can tell you how important that is. They are also the key ingredient in right-angled triangles, more of which in the next chapter.

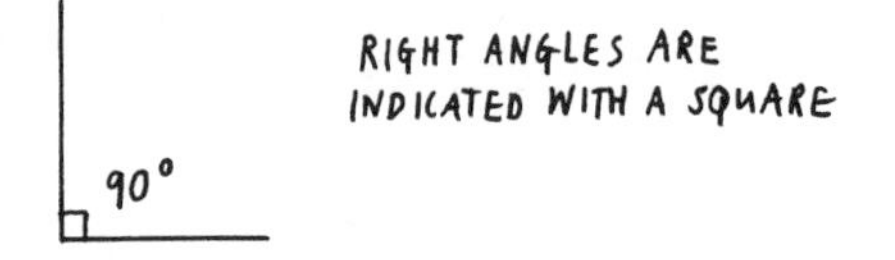

You can construct a right angle using just a compass and a straight edge. Start by drawing a line, which we'll call AB.

Take your compass, put the point in A, and draw a circle (it doesn't matter how large the circle is, so long as its diameter is longer than AB). Then without adjusting the compass, place the point in B and draw another circle.

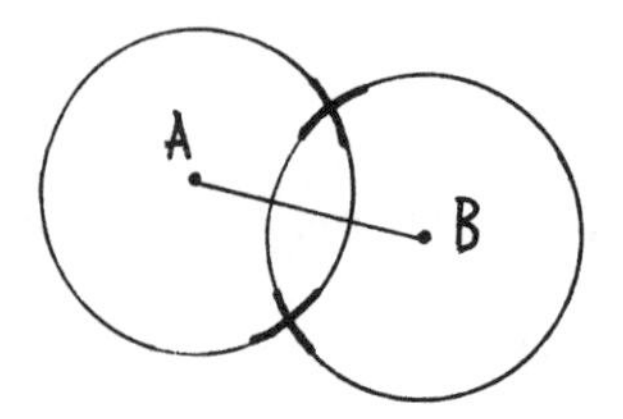

You don't need to draw the whole circles, the only important parts are the arcs where the two circles cross, which we have made bold.

Join the points where the two arcs meet with the straight edge to form the line CD.

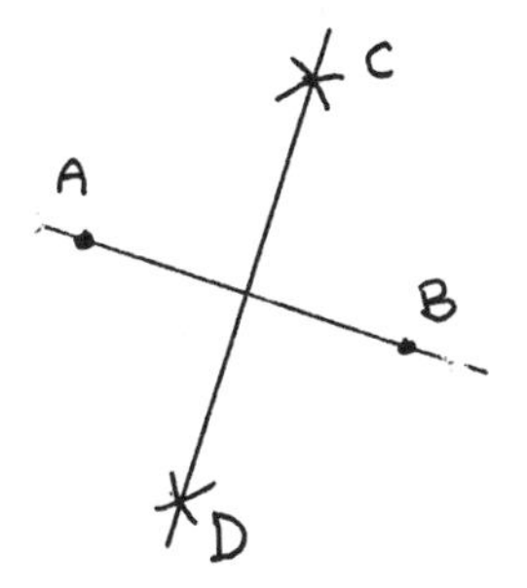

The two lines AB and CD are perpendicular, another way of saying that the angles where the lines meet are all 90°. CD is usually described as being the *perpendicular bisector* of the points AB. (Notice that perpendicular does not mean the same as horizontal or vertical – a mistake often made by teenagers.)

One other feature of a perpendicular bisector, obvious if you think about it, is that if you choose any point along it, that point is exactly the same distance from A as it is from B. The technical term for this is that the line CD is the *locus* of the points that are the same distance from A and B. 'Locus' in maths means 'the set of points that share a particular property', and is one of those words that previous generations of school children tended not to use.

TEST YOURSELF

Bob has a rectangular garden ABCD that is 8 metres wide and 4 metres long. A tree is in the corner near A, one metre from both fences (shown in the diagram). He asks a gardener to make a flower bed. The rules are that the flower bed should be all the points that are 2 metres or less from the tree, and are closer to AD than AB. Shade the locus of the flower bed.

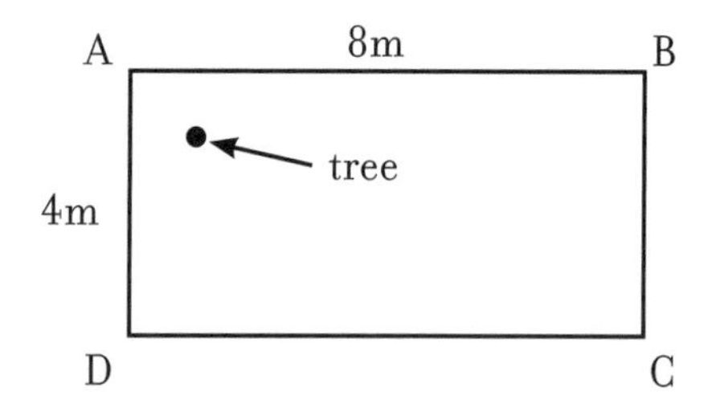

Interior Angles and Exterior Angles

There is another important set of angles in shapes, known as the interior and exterior angles. The 'interior' angles of a polygon are what you'd expect them to be, the angles on the inside of the shape. They are indicated on these two shapes below, the quadrilateral on the left and the hexagon on the right:

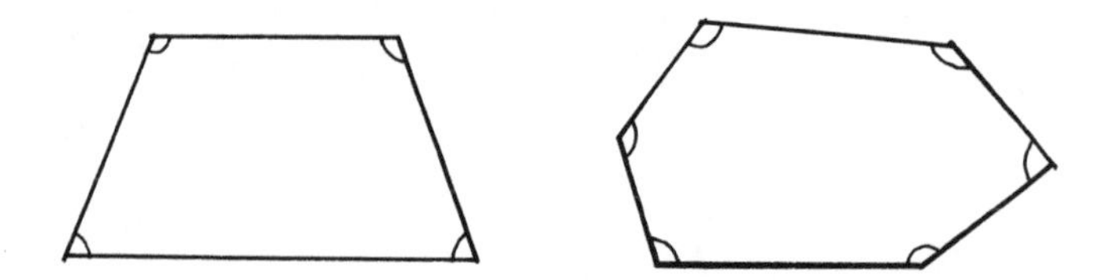

The 'exterior' angle is less obvious. Many people would expect the exterior angle to be the whole angle outside the corner (or *vertex)*

of a polygon, what we've sloppily called 'the outside bit' in this diagram:

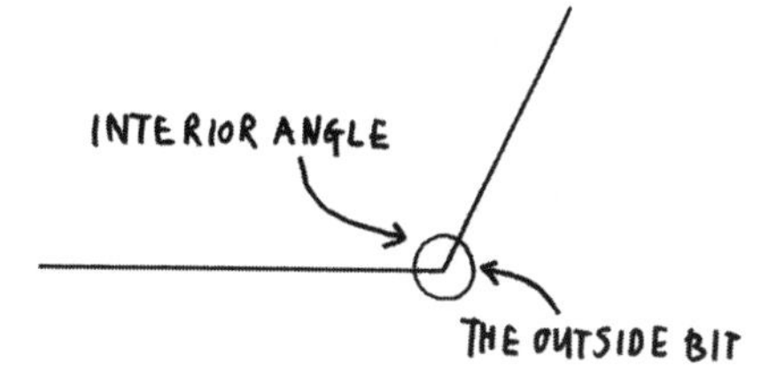

However, the convention in geometry has always been to define the exterior angle of a polygon as this:

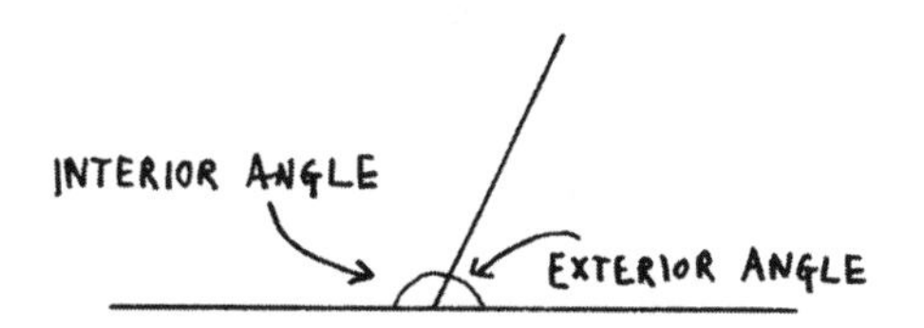

You can picture this as the polygon standing on the floor. The exterior angle is the angle between the outside of the polygon and the floor. This means that the interior angle and the exterior angle form a straight line and always add to 180°.

The exterior angles of a polygon also have the interesting property that they always add up to 360°. One way of seeing this is to imagine the polygon as being the walls of a building. As you walk along a wall of the building and come to a corner, the exterior angle is the angle by which you have to turn at that corner. If you do a complete circuit of the building you have done a full revolution, which means that the exterior angles must add up to 360°.

As for interior angles – we'll talk more about interior angles on page 160.

The hidden code of geometry diagrams

A good geometrical drawing will have sufficient markings that it tells you everything that you need to know about a particular shape. Teenagers might find it more interesting to think of these markings as a code to be deciphered, like the tracking symbols that are still used by Scouts. For example, here are two shapes, a triangle and a quadrilateral (a four-sided shape):

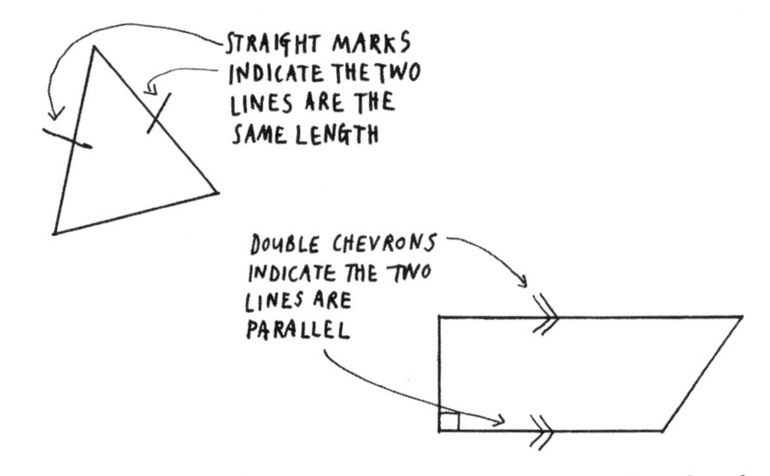

What do we know about them? The straight marks on the sides of the triangle indicate that those two sides are the same length – so it is an isosceles triangle. On the quadrilateral, the little square in the corner indicates that this is a right angle. Meanwhile, the arrows (more accurately the chevrons) on the two sides mean that those two sides are parallel. We therefore know that the top left angle is also a right angle, though we don't know anything about the other angles, or the lengths of the sides.

TEST YOURSELF

What do you know about this inaccurately drawn shape?

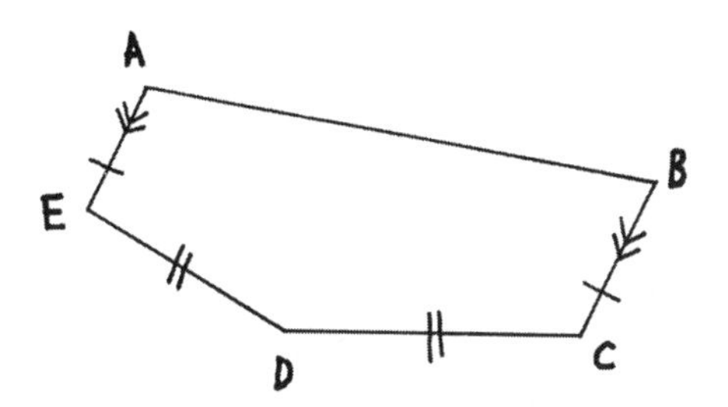

Circles, Triangles and Polygons

We have only begun to scratch the surface of the geometry of shapes. And yet, although proofs get more and more involved as you advance in geometry, most of the ideas follow on from the three constructions and proofs that you saw earlier in this chapter:

- creating equilateral triangles (page 143).
- the angles along a straight line (in the box on page 151).
- the perpendicular bisectors (page 153).

Indeed the vast majority of Euclid's geometry proofs are based around circles, triangles and perpendiculars, and the connections between them. You have to turn to his fourth book before you encounter the first pentagon, and hexagons and the other polygons get even less of a look-in.

Why didn't Euclid go big on pentagons, hexagons and the other polygons? The reason is that every straight-sided shape (polygon) can be reduced to triangles. So if you want to work out angles within a pentagon or more complicated shape, it can always be done by reducing that shape to triangles instead, and then simply using the maths of triangles.

In fact, triangles are the linchpin to so much maths that we have dedicated the whole of the next chapter to them.

THE PROPERTIES OF SHAPES

If you do only three things . . .

- Think of geometry as a logical puzzle to be solved.
- Look for real-world examples of parallel lines being crossed by other lines to form Zangles. They are surprisingly common: you'll see them on a garden trellis, a curtain draped across a blind, the grill on the front of most Volvos. Seeing that these appear in the physical world can make geometrical problems feel more real, too.
- Make your own pair of perpendicular lines or an equilateral triangle using just a pencil, a compass and a straight edge. It's quite a satisfying experience, and that satisfaction might even rub off on your teenager.

TRIANGLES

What do Buzz Lightyear and Lara Croft have in common? The answer is triangles. One of the most ingenious applications of geometry in the last decade has been its use by animators to help them create lifelike computer-generated images on screen. Zoom in on to these apparently bulging and curvaceous human forms and you will discover that their entire bodies are constructed from tiny triangles – and that is because a) triangles are the simplest shapes to deal with mathematically, and b) thankfully, all shapes can be broken down into triangles.

Why do these animations need to be 'dealt with mathematically'. The reason is that as the characters move, the angle and amount of light on each point needs to be calculated to mimic the appearance of real objects. The processing power of computers is so immense that the orientations, colours and relative positions of every triangle can be calculated almost instantaneously hundreds of times per minute, giving the impression that we are watching reality in motion, not mathematical shapes.

Although the maths involved in animation is often beyond A-level standard, at the heart of it are four topics that are first encountered before GCSE:

- The internal angles of a triangle
- The area of a triangle

- Pythagoras' theorem
- Trigonometry, in the form of sine, cosine and tangent.

Of course this isn't the only application of the maths of triangles, it crops up all over the place in design, navigation, engineering, space travel and elsewhere. The roofs and walls of modern buildings like the Eden Project domes in Cornwall and the Gherkin in London are filled with panes of glass that required complex triangle calculations. Triangles also happen to be the most 'stable' (unbendable) mathematical figures, which is why you'll see so many on them in pylons, climbing frames and other features of the built environment. But for many it's the Hollywood factor that most sparks the imagination. So if your teenager starts getting bogged down in the detail of what follows in this chapter just remember: *To infinity and beyond.*

Internal Angles of a Triangle

One fact that a teenager is expected to learn is that the internal angles of a triangle always add up to 180°, a piece of knowledge that is important in solving numerous geometrical problems. And most teenagers remember the fact thanks to endless repetition. However, not many teenagers know *why* it is true – and yet the proof is extremely simple, especially if they understand the opposite angles theorem (the one we put in a box on page 151).

Take any triangle with internal angles a, b and c, and stand it on the floor on its longest side:

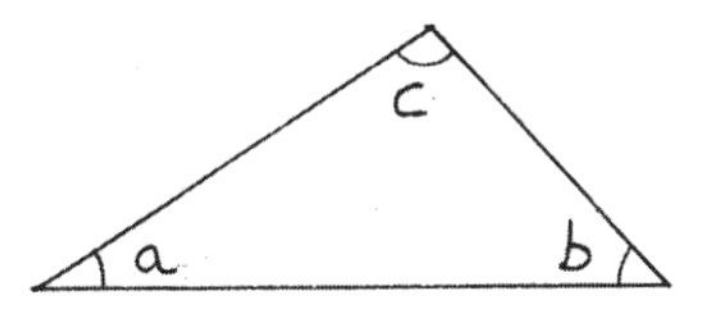

Now imagine that this triangle touches the ceiling (the floor and ceiling are parallel, of course):

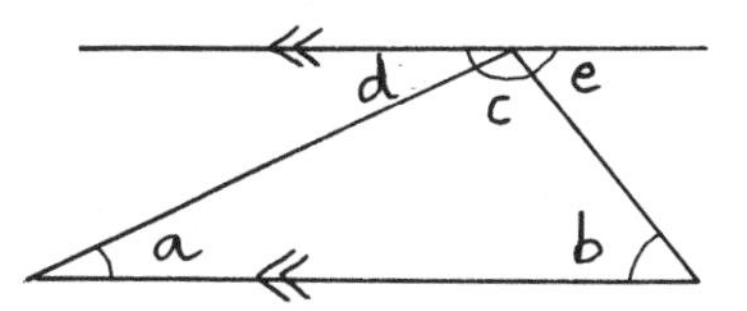

Since angles c, d and e form a straight line, they must add to 180°. But notice that angles d and a form a Zangle, and we discovered on page 148, that the angles in a Zangle are always equal.*

In the same way, angles e and b form a (backwards) Zangle. That means d is the same as a, while e is the same as b, so we can replace d with a, and replace e with b. In other words, the three angles $c + d + e$ are the same as $a + b + c$.

And since $c + d + e = 180$, *that means* $a + b + c = 180$. QED (which stands for *quod erat demonstrandum*, 'that is what we wanted to prove').

There's a neat way to demonstrate this, by the way. Draw a triangle on a piece of paper, then tear off the three corners. When you fit the corners together, they will always form a nice straight line.

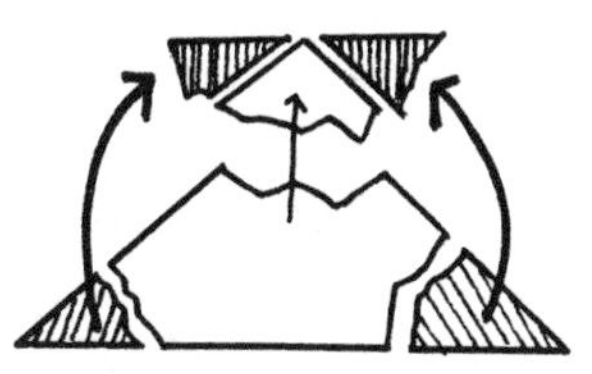

* This sounds like something from Dr Seuss.

This becomes useful if you are only told what two of the angles in a triangle are: it will always be possible to work out the third, because the three angles together must add to 180°. And sometimes, if it's a particular sort of triangle, then you only need to know one angle. For example, in the isosceles triangle below, angle *b* must also be 50°, so angle *c* must be 80°.

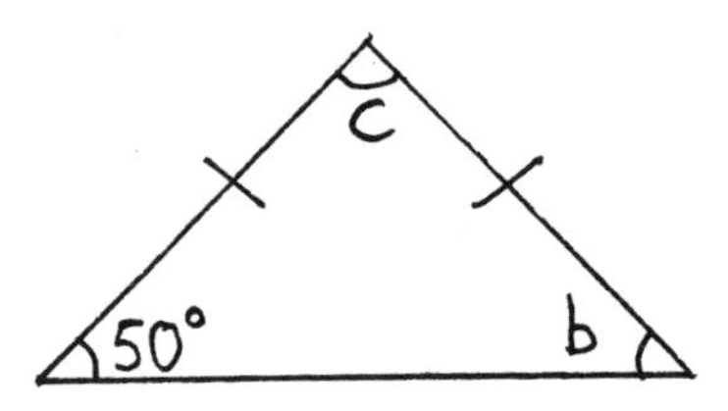

How's your geography?

Comedian Milton Jones included this triangle geometry in his one-man show.

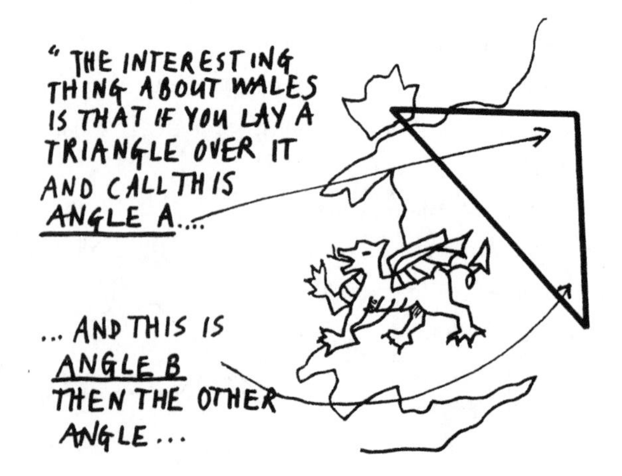

Interior Angles of a Polygon

Knowing that the interior angles of a triangle add to 180° makes it easy to work out the interior angles for *any* polygon, because any shape can be divided into triangles. What do the interior angles in these two shapes add up to, for example?

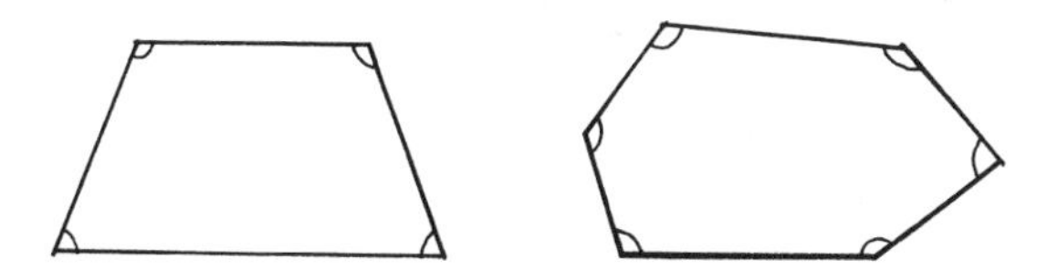

It looks difficult, because these aren't regular symmetrical shapes. But in each case we can divide the shape up into triangles. The shape on the left, a quadrilateral, can be divided into two triangles. Since the internal angles of each triangle add up to 180°, the combined angles of both triangles must add to 360°. And this will be true for any quadrilateral. The simplest example of a quadrilateral is a square, and sure enough, a square has four right angles, and $4 \times 90 = 360$.

The shape on the right, an irregular hexagon, can be broken up into four triangles, so its interior angles add to $180 \times 4 = 720°$. All hexagons can be divided into exactly four triangles, and hence the interior angles of all hexagons will be 720°.

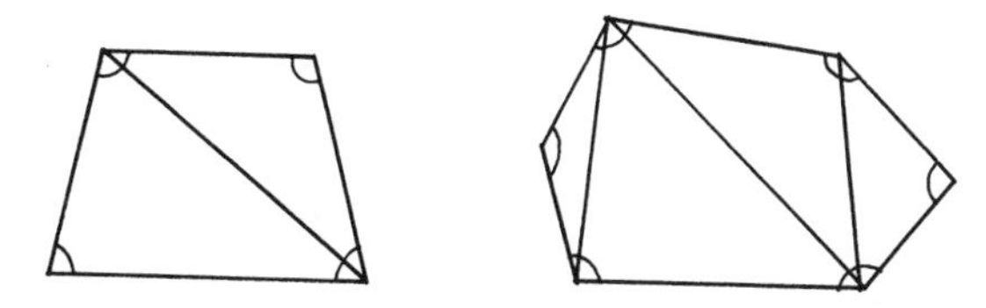

In fact, you can easily test out for yourself that a polygon with N sides can be divided into N-2 triangles (as long as the lines you use to divide the polygon up don't cross each other).

The results are set out in the table below, including the interior angles of the *regular* polygons (those whose sides and angles are all the same).

Polygon	Number of sides	Number of triangles it can be divided into (T)	Total of the interior angles, A (T × 180)	Interior angle in the regular polygon
Triangle	3	1	180	180÷3 = 60°
Quadrilateral	4	2	360	360÷4 = 90°
Pentagon	5	3	540	540÷5 = 108°
Hexagon	6	4	720	720÷6 = 120°

. . . and as a general rule an N-sided regular polygon will have an interior angle of $180(N - 2) \div N$.

TEST YOURSELF

Here is an extremely irregular 12-gon (*dodecagon*). What do its interior angles add up to?

And what is the interior angle of a *regular* dodecagon?

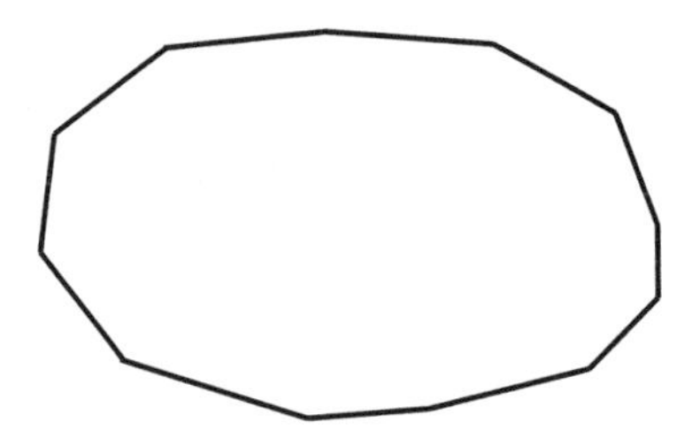

The Area of a Triangle

Another property of triangles that crops up frequently is the formula for finding its area.

The area of a triangle is half the length of the base multiplied by the height, or: $A = \frac{1}{2} b \times h$.

Most teenagers memorise it (or look it up on the formulae page) without understanding why it works, but the explanation of it is very simple. Take any triangle and rotate it so that its longest side is the base. We'll call the length of the base b:

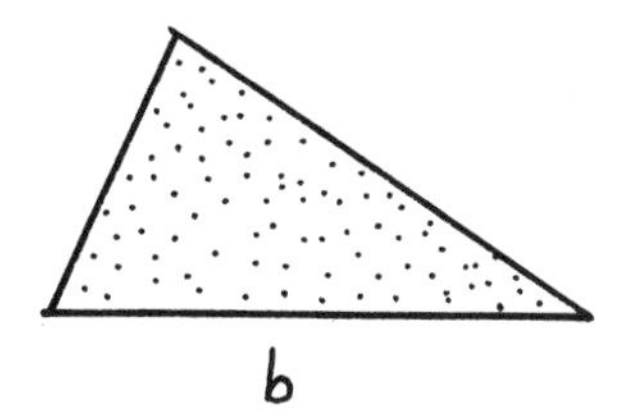

Now imagine the triangle fitting snugly into a rectangle, whose base is b and height is h.

The area of the rectangle is simply its base × height, and the diagram shows that the triangle is half of the rectangle, so its area will be half of the rectangle's, or $\frac{1}{2} b \times h$.

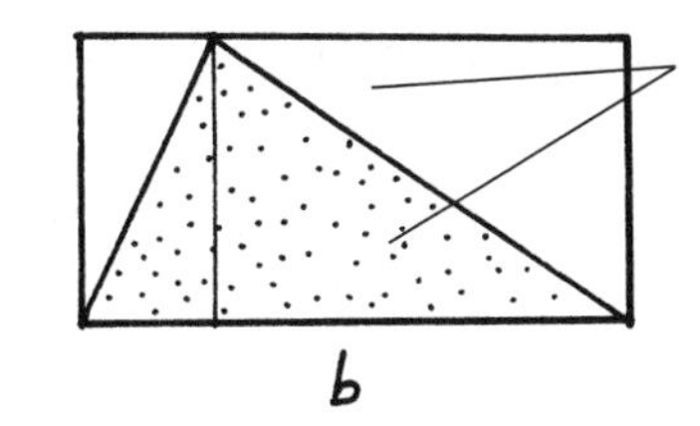

This shaded triangle and the white triangle are identical, and the same is true for the smaller shaded and white triangles, so the large triangle must be half the area of the large rectangle.

Finding the Missing Information

We've seen that if you know two angles of a triangle then you can immediately work out the third (because the three add up to 180).

So intimately connected are the sides and the angles of a triangle that you need surprisingly little information about the triangle in order to reproduce it exactly. Each of the following is enough to precisely define the dimensions and angles of a triangle:

- The length of all three sides, but none of the angles.
- Two angles and the length of the side that connects them.
- The length of two sides and the angle between them.

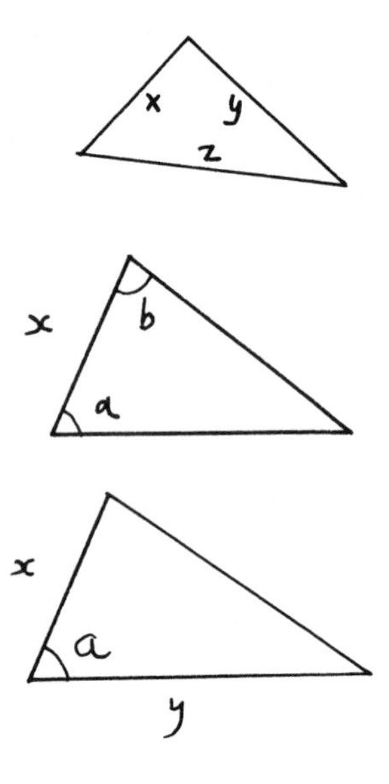

There is lots of maths that uses this knowledge to work out the missing information. We'll look at two of the most important that your teenager will encounter: the trigonometry of sine, cosine and tangents; and first, the theorem of Pythagoras.

Pythagoras' Theorem

Pythagoras was a mathematician and mystic in ancient Greece. He's most famous for a theorem about triangles that he didn't actually discover – the Babylonians knew about it centuries earlier.

Of all the geometry learned at school, Pythagoras' theorem is the one that has the most practical, everyday use. Most often it is used to work out the straight line distance between two points. Here, for example, is an anecdote from one parent that we met:

'Early in my career I had to write a computer program for helping to decide the best locations for mobile phone masts. The program needed to know the distance between two neighbouring masts so it could work out how strong the phone signal would be. I had the co-ordinates of each mast (east and north) and realised I could picture these two masts as being the corners of a right-angled triangle. To work out the distance from one mast to the other I programmed the computer to use Pythagoras to work out the hypotenuse. I remember feeling very pleased that I'd used a bit of school maths to crack the problem.'

Here is a simple reminder of the theorem.

> *In any right-angled triangle with sides of length* a, b *and* c *(where* c *is the longest side, the hypotenuse) then (as we'll demonstrate shortly) it will always be true that:* $a^2 + b^2 = c^2$

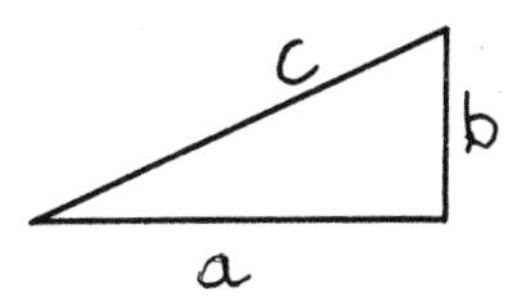

What this means is that if you know a and b (the lengths of two sides of a field, say), you can use the formula to work out c (the length of the diagonal path across the field).

It is true for any other combination of sides too, so if you know a and c, you can figure out b.

Take this example:

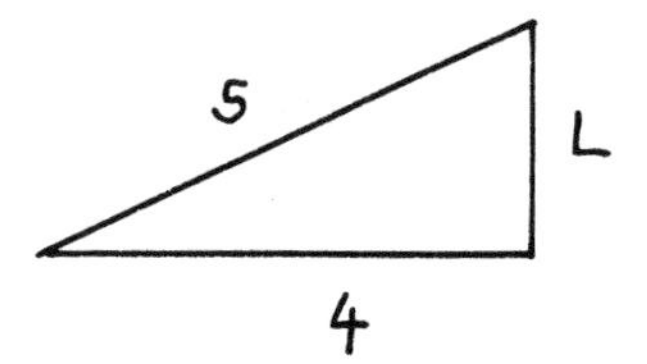

What is the length of the missing side, L? We know that $4^2 + L^2 = 5^2$, so $L^2 = 5^2 - 4^2 = 25 - 16 = 9$.

So in this case L is the *square root* of 9, which is 3. The missing side doesn't always work out to be a nice whole number like this, but what we have here is a 3–4–5 triangle. The 3–4–5 right-angled triangle has important historical significance because it was known to the ancient Babylonians and Egyptians, and shows that the so-called theorem of Pythagoras was known long before this mystical Greek mathematician had his name linked to it. There are stories that the Egyptians used to use a piece of rope with twelve equally spaced knots, and form the rope into a triangle with four spaces between knots on the base and three up the side and five along the diagonal to form a do-it-yourself right angle. This would have been a handy device to have in your toolbox if you were wanting to make the corner of, say, a pyramid. Whether the Egyptian story is true or not, the twelve-link chain has certainly been used by surveyors in more modern times.

Some teenagers can understandably get confused about whether Pythagoras is about lines or squares. It's actually about both. The area of a square is the length of its side 'squared', so what the theorem is saying is that for a right-angled triangle, the area of square C = Area of square A + Area of square B:

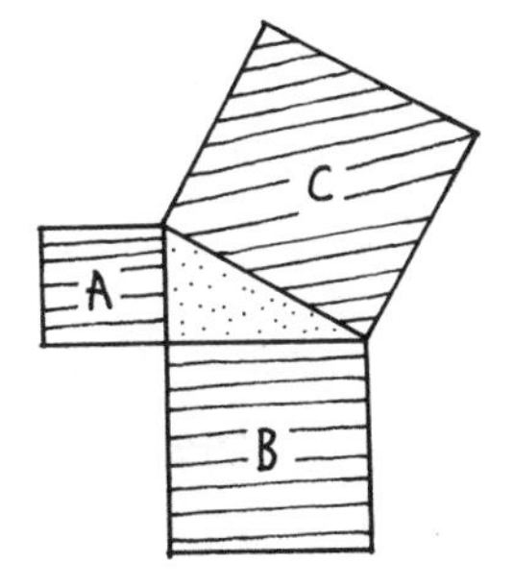

Pythagoras' theorem is usually proved by cutting up the areas of squares to show that they are equal. This has been of such fascination to mathematicians over the centuries that there are known to be well over 300 different proofs of it. Our old friend Euclid (see page 140) came up with a proof, but there have been more modern proofs too, including, rather bizarrely, one by a nineteenth-century American president, James Garfield.

Although teenagers are not expected to prove Pythagoras, there is a lovely proof that can be done by sliding pieces of paper. Copy square C from the diagram above, and surround it with a larger square. Now make four copies of the original triangle out of paper – these four copies will fit perfectly into the gaps around square C.

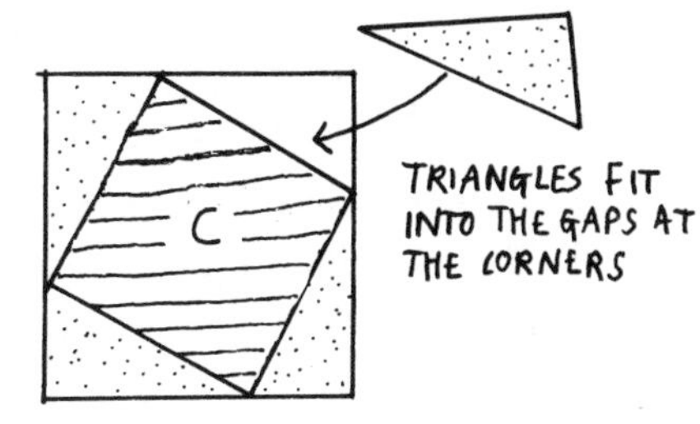

Think of the area C as being bare floorboards, while the triangles are pieces of carpet. Now slide the top right triangle (carpet) down to the bottom left, and the top left and bottom right triangles across to join each other in the top right corner, as in the diagram below. The gaps that are left form two squares, one with side A and the other with side B. These are the new areas of bare floorboards, and their combined area must be the same as the original area, C. In other words, the area occupied by C is the same as the one occupied by $A + B$. That's Pythagoras' theorem.

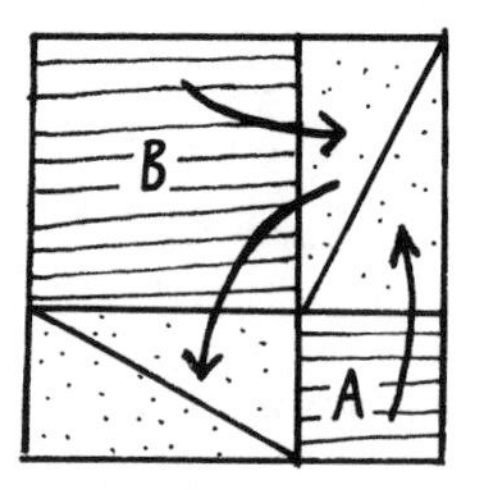

The ratios of triangles' sides

Pythagoras is an important tool for working out missing information in a triangle, but it belongs to a wider topic known as trigonometry. Trigonometry is another word that comes from Greek. It means 'the measurement of triangles'.

The key idea behind trigonometry is simple. If two triangles have the same angles, but one triangle is an enlarged version of the other, then the ratio of the length of their sides will always be the same. For example, triangles ABC and ADE are *similar* (i.e. their angles are the same).

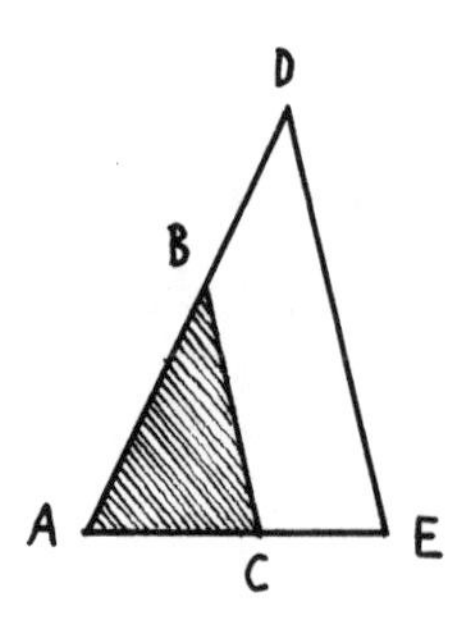

This means that the ratio of AC to AE will always be the same as the ratio of AB to AD.

On page 48 we talked about how being able to work out ratios is one of the vital skills in maths. Trigonometry is another example of where ratios crop up, and where teenagers regularly make mistakes. For example, in the triangle above, if AE is 3cm longer than AC, many teenagers will think that AD is also 3cm longer than AB.

TEST YOURSELF

In the triangle on page 170, suppose the length AC is 5cm, AE is 8cm and AB is 7cm. What is the length of AD?

Sine, Cosine and Tan(gent)

What happens if one of the angles in a triangle is a right angle? For right-angled triangles, you only need to know *two* pieces of information (one angle, call it x, and the *length* of one side, or alternatively the *lengths* of *two* sides) in order to work out all the other angles and side lengths. The three sides of a right-angled triangle each have a name.

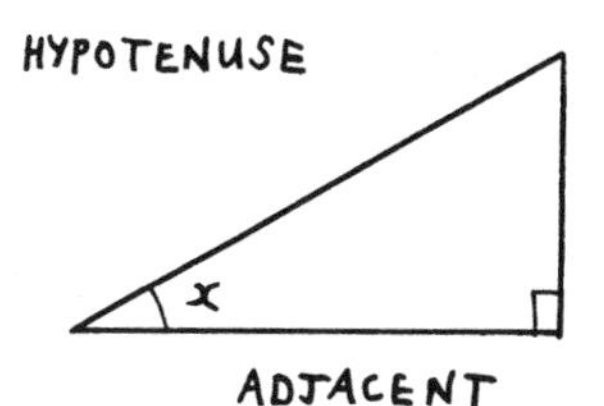

The longest side is the *Hypotenuse* (this word literally means 'under stretch' – perhaps it resembles a piece of string like a washing line, stretched between the end of a vertical pole and the ground). The side next to the angle x is called *Adjacent.* And the side opposite x is called, unsurprisingly, *Opposite.*

In the three right-angled triangles below, the angles are all the same. Although the triangles are different sizes, the *ratios* of the sides to each other will always be the same, and, very helpfully, mathematicians have worked out all of these ratios so that you don't have to. School children used to have to look these ratios up in books of tables, but now there is a calculator button that takes away all the strain.

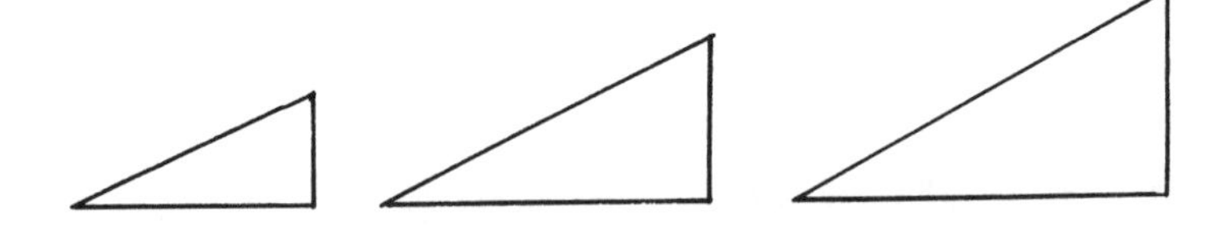

The ratios between the sides all have names:

- Opposite/Hypotenuse is called the *sine* (usually shortened to *sin* but still pronounced so that it rhymes with wine).
- Adjacent/Hypotenuse is called the *cosine* (shortened to *cos*).
- Opposite/Adjacent is called *tangent* (shortened to *tan*).

And thanks to buttons on a scientific calculator, if you know two sides of a right-angled triangle, or one side and one angle, you can use sine, cosine and tangent to work out, respectively, both of the other angles or the lengths of the other sides of the triangle.

Example 1: you know the lengths of two sides

Here's a triangle with one side 5cm and the hypotenuse of 8cm. What are the two angles, *a* and *b*?

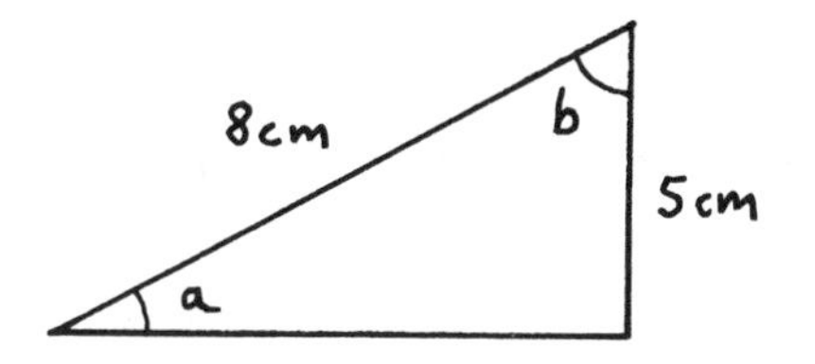

- Opposite ÷ Hypotenuse = 5 ÷ 8 = 0.625 = sin (*a*). We need to know what 0.625 is the sine *of*, which means using the *inverse* button on your calculator. The inverse of a trigonometry function is usually indicated like this: $\cos^{-1}$ or $\sin^{-1}$ or $\tan^{-1}$. How you actually *do* $\sin^{-1}$ (0.625) depends on your calculator – it might involve pressing an INV or SHIFT key. Whichever it is, you should end up with the answer 38.7° (rounded to one decimal place).
- Now that we know *a*, we can work out *b* because the three angles of the triangle $a + b + 90$ add up to 180°. This means $b = 51.3°$.

Example 2: you know one side and one angle

In this triangle we know one angle is 30^0 and the opposite side is 10cm.

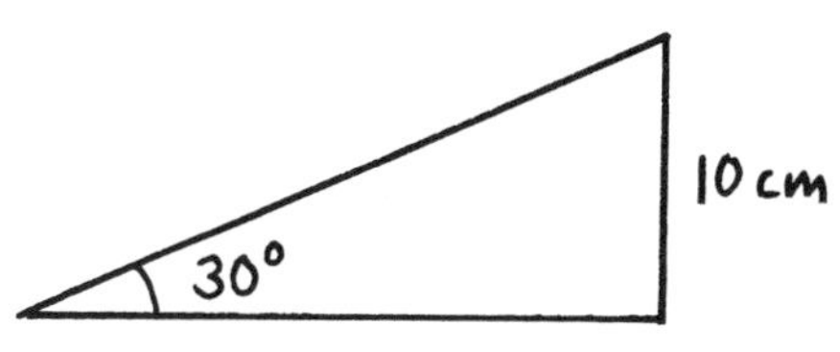

- 10 ÷ Hypotenuse = sin 30 (which your calculator will reveal is 0.5), so the hypotenuse must be 20cm.
- 10 ÷ Adjacent = tan 30 (roughly 0.58) so the base of the triangle is 17.32cm.

The problem teenagers have is actually applying this in practice. Two issues typically crop up:

1) Remembering which one is sine and which is cosine.
2) Rearranging the equations and using a calculator properly to work out the answer.

Generations of teenagers have been taught a mnemonic for the first of these, namely:

SOHCAHTOA (which stands for Sine = Opposite ÷ Hypotenuse, Cosine=Adjacent ÷ Hypotenuse, Tangent = Opposite ÷ Adjacent).

Another way to remember it is to think of cosine as the one where the sides are 'cosy' with each other, snuggling up around the angle.

There are plenty of other mnemonics out there for the same thing, including some more risqué ones like: 'Thousands Of Abortions Come After Having Sex On Holiday'.

As with all these things, however, given enough practice you just remember which one is which.

TEST YOURSELF

You will need a calculator with sin(e) and cos(ine) buttons to work this out.

a) If the angle at corner A (usually denoted as angle BAC) is 40°, and the length BC is 10cm, what is the length of AB?

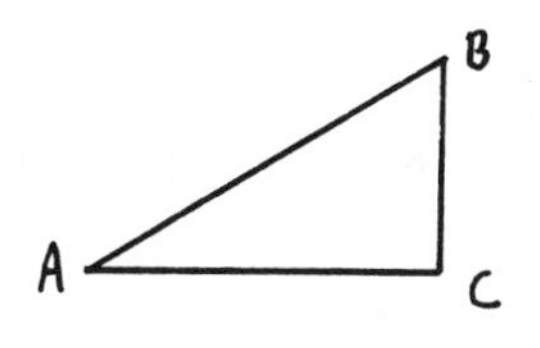

b) What is angle ABC in this triangle (i.e., the one in the top right) if AB = 200cm and BC = 25cm?

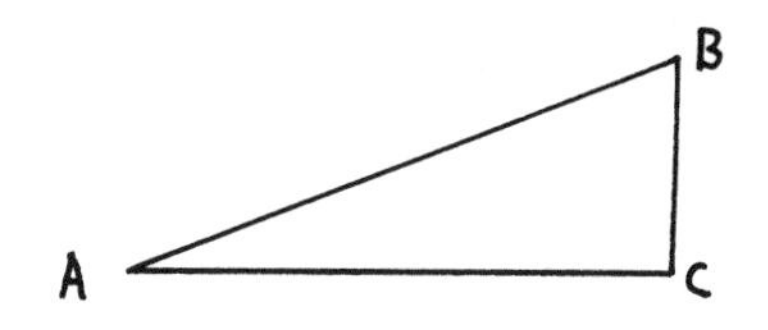

TRIANGLES

If you do only three things . . .

- Search the Internet for some close-up images from animations and games so you can demonstrate how even the most complicated curved surfaces are built up from triangles.
- Put up Post-its on the fridge with trigonometry rules to remember (angles of a triangle, the formula for area of triangles, Pythagoras) – if your teenager regularly comes across these they will begin to become second nature. (But make sure you occasionally change the facts that are on display.)
- Encourage your teenager to scribble all over homework sheets – mark the parallel lines on diagrams, which angles are equal to other angles, and so on. Doing this before immediately trying to answer a question can often reveal easy solution methods.

MOVING AND CHANGING SHAPES

For anyone with good visualising skills, the maths of moving and changing shapes (by sliding, rotating, and reflecting them) can be great fun. Doing *transformations*, as this is called, can even be artistic. The fantastically intricate mosaics in medieval mosques, the dramatic roofline of Sydney's famous Opera House, and the realistic impression that you're flying when you sit in a flight simulator are all thanks to people who understood precisely what happens to shapes when you move and change them.

Alas, the gift of visualising does not come naturally to everyone, and for many teenagers, this can prove to be a surprisingly tricky area of maths.

There are two types of transformation. The first type are those in which the shape changes position but stays the same size (sliding, rotation and reflection). These are typically the ones used by designers of textiles, wallpaper and anything else with a repeating pattern.

The second type are those transformations where the size or the shape changes. These include enlarging, shrinking and stretching.

Sliding or Translating

More formally this is called translating, the shape moves position but finishes 'the same way up':

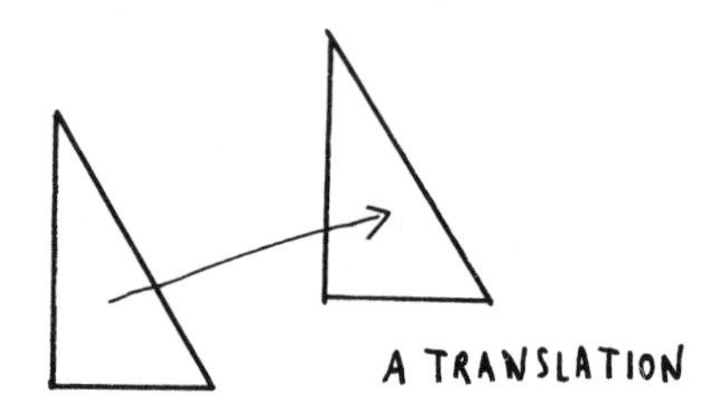

A *translation* is usually described in terms of the distance along and up, and written in the form of a *vector* (see page 191). It's also possible to describe it in terms of the distance moved and the angle at which it moves, what are known as *polar co-ordinates* (but it's extremely rare for these to be taught before A level).

Reflection

Exactly as it sounds, this is what would happen to the shape if reflected in a mirror. With a reflection, the further that a part of the shape is from the mirror, the further it will 'travel':

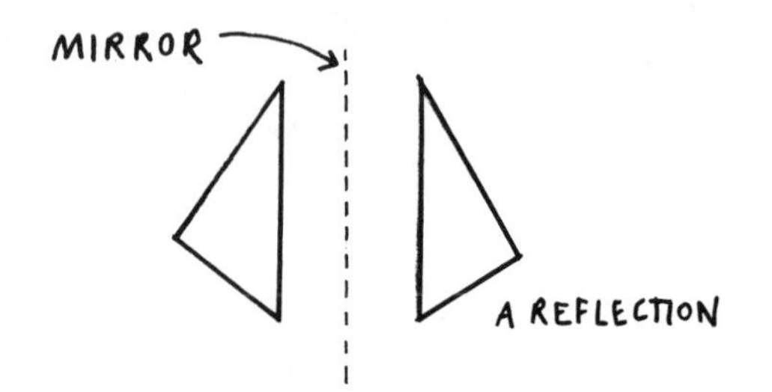

Most teenagers have little difficulty picturing or drawing a reflection when the mirror is vertical. It's when the mirror is at an angle that the problems arise.

Here's an example of a typical error:

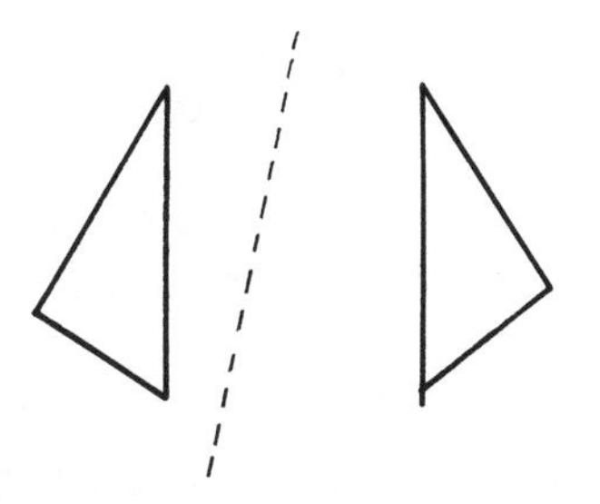

The simplest trick for visualising the right reflection is to rotate the paper until the mirror line is vertical or horizontal, and then to draw lines from key points on the shape that meet the mirror at right angles (i.e., are perpendicular to the line). The reflected points of the shape line up along that line, the same distance from the mirror.

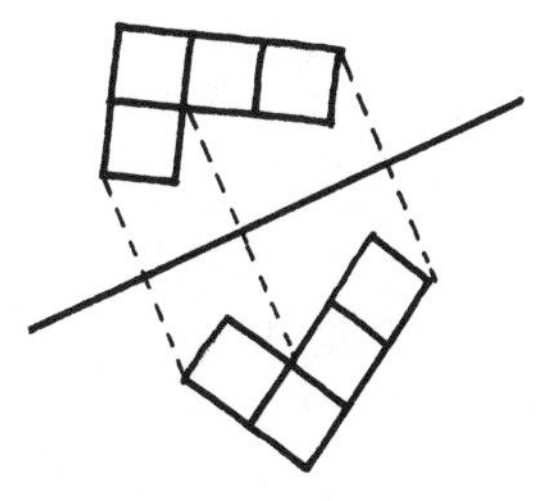

Tracing paper can help in understanding reflections. (We know adult designers who still use this technique as a check – if they don't have access to a computer package that will do it for them.) Put the edge of the tracing paper along the mirror line, trace the shape and then flip the paper over.

Rotation

Rotation is simple when a shape is rotated around its centre. Here's a square that has been rotated by 45° around the point X:

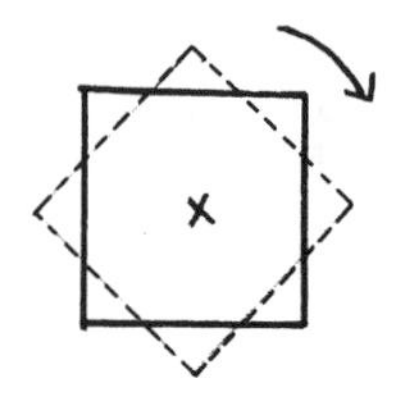

But some rotations are harder to figure out. In the next example the triangle has been rotated clockwise by 90°, but the point that it has been rotated about is not inside the triangle. The centre of rotation has been indicated with an X.

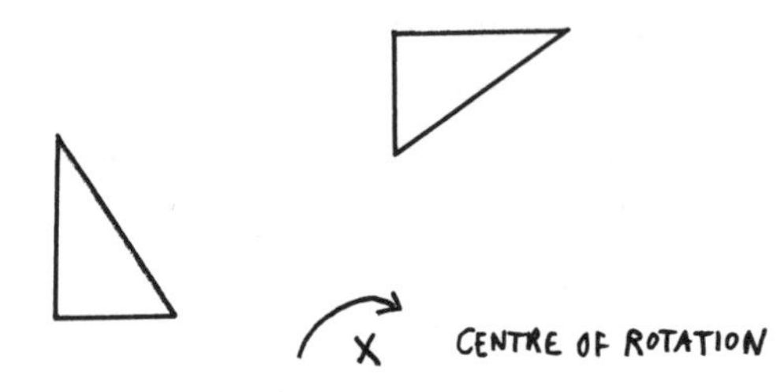

The centre of rotation can be easier to understand by relating it to the real world. A child sitting on a carousel horse at the fairground is rotating, and the centre of rotation is the hub of the carousel. If you imagine the triangles above as being a horse moving around the carousel, with X at the hub, it's easier to imagine what is going on.

Again, tracing paper can help, this time with the aid of a drawing pin. Draw the shape to be rotated and place the drawing pin at the centre of rotation. If the challenge is to *find* the centre of rotation, some trial and error with the drawing pin's position will quickly locate it for you.

Finding the centre of rotation using geometry

There's a way to find the centre of rotation using geometry. In fact, it uses the idea of perpendicular bisectors that we met on page 153. This is much more elegant than using tracing paper, though it requires a bit more thought.

Below is a triangle that has been rotated. Where is its centre of rotation? Remember the analogy with the fairground carousel. When moving around the carousel, the distance of everything from the hub remains the same. So every point on the triangle must always remain the same distance from the centre of rotation. Pick any point on the triangle, for example the corner A that forms a right angle, and join the corresponding points on the 'before' and 'after' triangles with a line. The centre of rotation must lie along the perpendicular bisector, shown as a dotted line, because those are the points that are equidistant from the corner of the original and its rotated image.

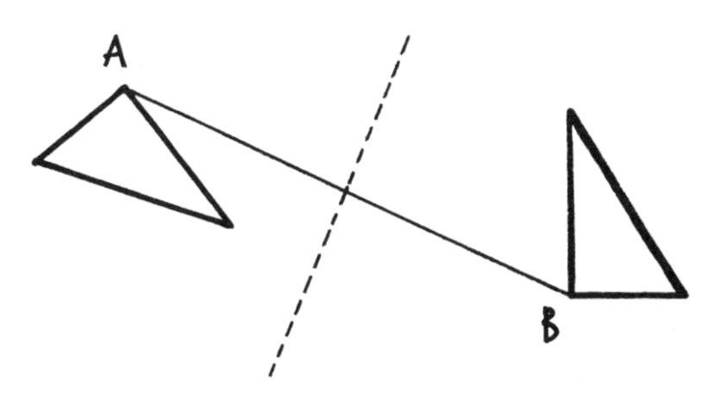

Next join any two other equivalent points on the triangle and find their perpendicular bisector. The centre of rotation must lie on both of these dotted lines, so the point where the (dotted) lines cross must be the centre of rotation.

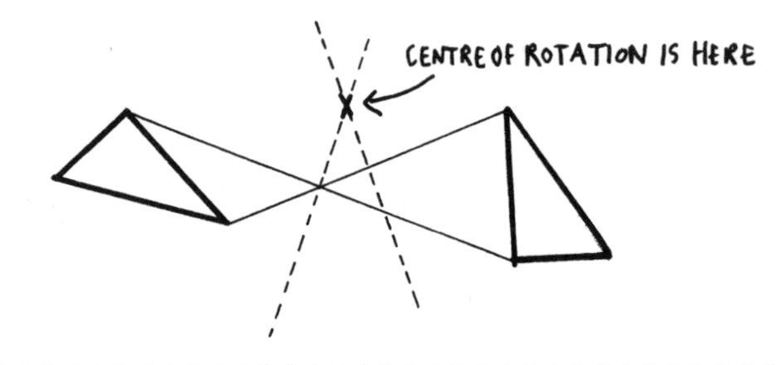

Wallpaper and Fabric Patterns

Reflection, translation and rotation are central to the designs that we use to decorate our world. Take a look at any wallpaper or tiling pattern and you'll see what we mean. The reason for this is a simple, practical one: because the shapes don't change as they are transformed, they can be produced by a simple printing block.

Any budding textile or graphic designer might enjoy the challenge of coming up with a completely new, repeating pattern. Although there are an infinite number of ways that a printing block can be designed, it has been proved that there are, in essence, only seventeen different wallpaper patterns using a combination of sliding, reflecting and rotating. Almost all of them can be found decorating the walls and floors of the spectacular Alhambra Palace in Spain. If your teenager is able to come up with an eighteenth, they'd not only be making history, they would be defying mathematics too.

Enlargement and Shrinking

So far we have looked at translations, reflections and rotations, none of which change the size or shape of the object. The second type of transformation is one where *the shape changes so that it no longer perfectly matches the original* – if you laid one on top of the other, there would be overlaps. The only transformation of this type that your teenager is likely to encounter at school is enlargement (and its reverse, shrinking).

Enlargement is what some photocopiers do. The size of the shape changes, but all of its angles and curves stay the same: a square stays a square, a circle is still a circle. A regular pentagon is still a *regular* pentagon when it is enlarged:

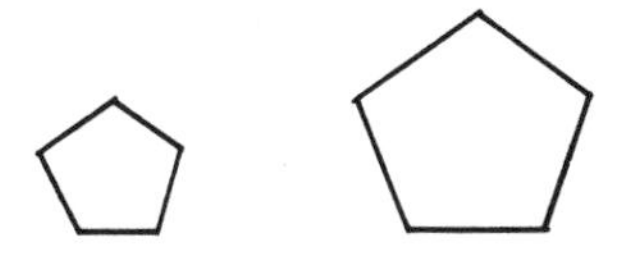

The amount by which a shape is enlarged or shrunk is called the scale factor: if a side of the original shape is 3cm and the enlarged shape is 12cm, that is a scale factor of four, and every side will be four times longer.* If a shape is made smaller then the scale factor will be less than one. The smaller shape that was scaled up by a factor of four can be regained by scaling the enlarged shape by a factor of $\frac{1}{4}$.

Just as a rotated object has a centre of rotation (we used the analogy of hub at the centre of a fairground ride), an enlarged object has a *centre of enlargement*. This point is easy to find,

* Note that this scaling only applies to lengths. One of the things that can confuse is the effect of scaling on area: a scale factor of 4 will make all lengths four times longer, but will make the area *16* times larger ($= 4 \times 4$, the square of the scale factor).

simply by drawing straight lines through equivalent points on the original and enlarged object. Notice that you can continue drawing the lines through the centre of enlargement and beyond. This creates inverted, or 'negative', enlargements. The flipped pentagon shown on the left is the same size as the original, enlarged with a scale of −1!

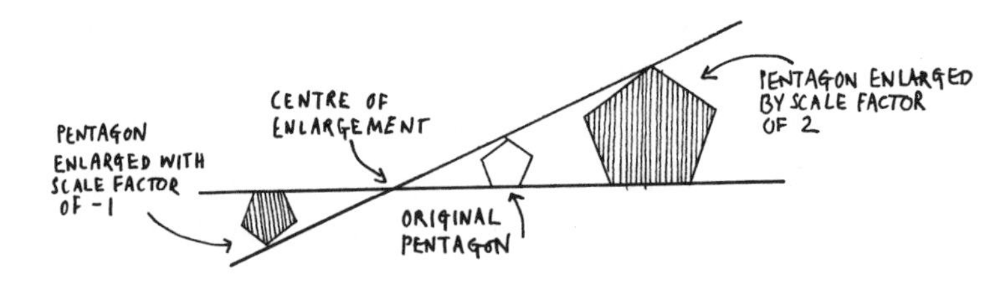

Stretching

Stretching happens in only one direction, like this, for example:

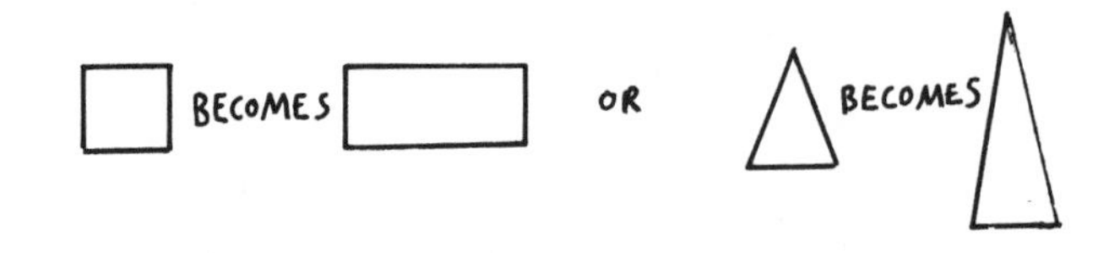

Notice how the shape becomes taller or wider, but not both. This means mathematical stretching isn't quite the same as stretching elastic (say) because if you stretch most elastic it gets longer *but it also gets thinner.*

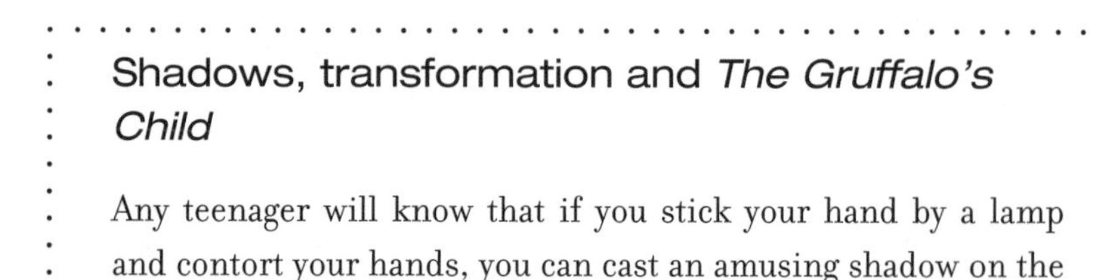

Shadows, transformation and *The Gruffalo's Child*

Any teenager will know that if you stick your hand by a lamp and contort your hands, you can cast an amusing shadow on the wall. Expert hand-shadowers can produce dogs, birds, and also a variety of rather ruder images. These shadows are enlargements. If your hand is midway between the lamp and the wall, the enlargement has a scale factor of two. The closer you move your hand to the lamp the greater the enlargement will be:

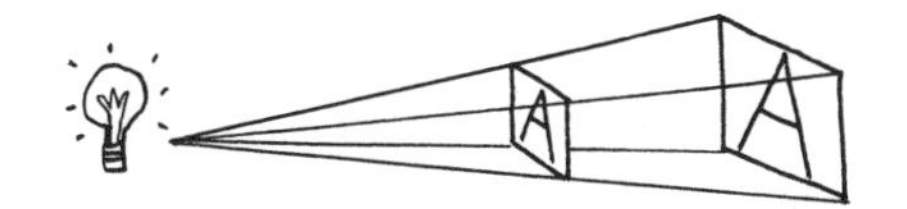

On the other hand, if the source of light is a long way away – the sun or the moon for example – then there will be almost no enlargement of the shadow at all, because the rays of light are (as near as makes no difference) parallel. This might seem surprising, but if you hold up your hand square to the sun and put a piece of paper behind it, also square to the sun, the shadow hand will be normal size. However, if you *tilt* the paper, the shadow will become stretched and you can create a long hand with claw-like fingers.

When your child was younger you might well have read them *The Gruffalo's Child.*

The story hinges on a tiny mouse convincing the monster Gruffalo that a giant scary mouse is nearby, by standing on a branch and letting the bright moonlight cast a giant mouse shadow. Now we really hate to be spoilsports, it is a charming story, but as we've seen above, the sun (or the moon) low in the sky doesn't enlarge your shadow, it merely stretches it. So what the Gruffalo's child should have seen was an incredibly long and skinny, 'normal width' mouse. And that might not have sent him running away in quite the way as happens in the story.

Symmetry

Symmetry is a subject children first encounter at primary school. The main difference at secondary school is that it becomes a little more formal, and the shapes being dealt with could be more complicated.

There are two sorts of symmetry that your teenager will be learning about – reflective symmetry and rotational symmetry – and they are closely linked to the transformations described earlier in this chapter.

Reflective symmetry

Reflective symmetry is the symmetry of butterflies, cartoon faces and curtained windows. We are naturally drawn to things with reflective symmetry, although even the most beautiful faces are not purely symmetrical. In fact, psychologists have shown that people prefer faces that are a bit 'flawed' to perfectly symmetrical ones. If you, or your teenager, are skilled in digital image manipulation then creating family portraits composed of the left side of the face reflected to create a full, perfectly symmetrical face (and similarly with the right side of the face) provides some interesting insights into what each side of your face reveals.

A common mistake teenagers make is thinking that a shape that can be cut into two matching parts will always have reflective symmetry. Cutting along the diagonals in each shape below would produce two identical (congruent) triangles, but the diagonal is only a line (or axis) of symmetry on the square. If you fold the rectangle or parallelogram along the diagonal it creates an overlap and so the diagonal is *not* a line of symmetry.

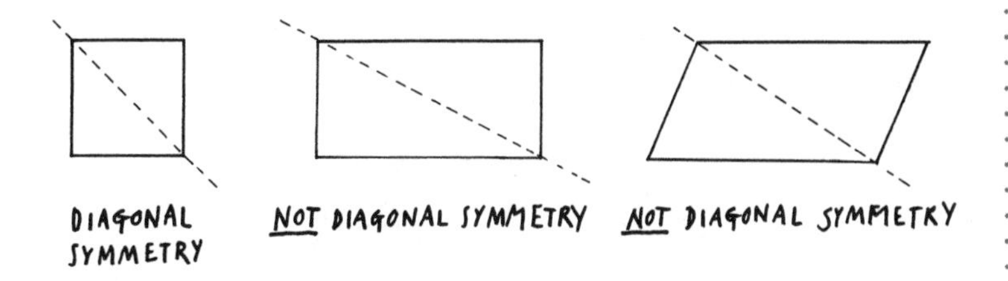

Shapes can have more than one axis of symmetry. The number of axes can be worked out by counting the number of lines along which a mirror could be placed to create the original shape. Rectangles have two axes of symmetry, while squares have four. Sloping parallelograms have none.

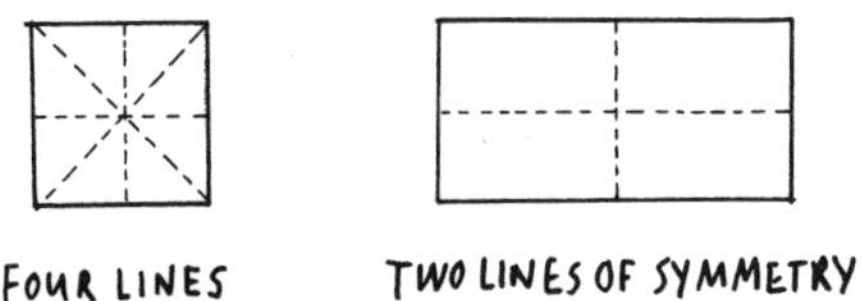

It is less likely that your teenager will have done much investigation of reflective symmetry of three-dimensional shapes. Just as in two dimensions, the axis of symmetry is where a mirror would be placed, in three dimensions a plane of symmetry is where a mirror could slice through an object and the combination of the part seen and the part in the mirror look exactly like the original object.

TEST YOURSELF

Here are two planes of symmetry on a cube. It's quite challenge to try to picture them all: what can feel like a different plane of symmetry can turn out to be one you've already put in place from a different position. A cube has nine planes of symmetry: can you picture them all?

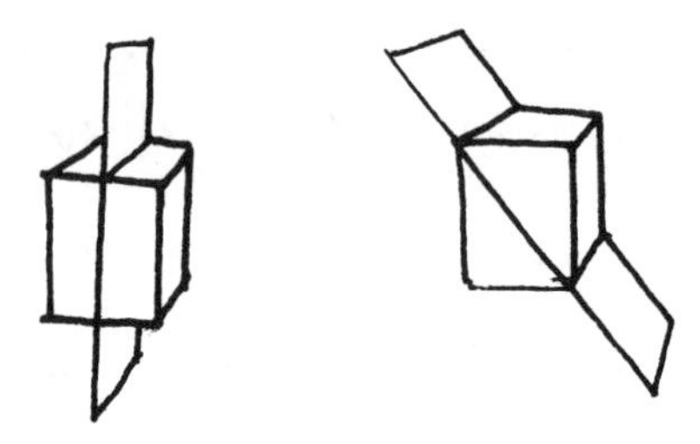

Rotational symmetry

Rotational symmetry is the symmetry of shapes 'looking the same' when they are rotated. Although not so obvious as mirror symmetry, we are also naturally drawn to shapes with rotational symmetry. For example the Isle of Man emblem has rotational symmetry.

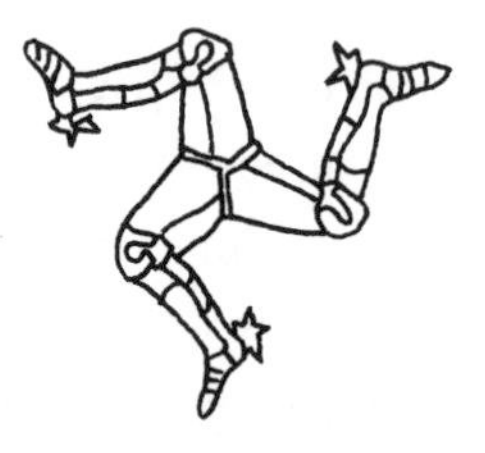

THE ISLE OF MAN LOGO HAS ROTATIONAL SYMMETRY OF ORDER 3

The informal way to think about rotational symmetry is to imagine drawing around the shape in question – is it possible to rotate the drawing so that it fits back over the original? The number of ways that this can be done is called the order of rotational symmetry. For example, an equilateral triangle has rotational symmetry of order 3: there are three ways any equilateral triangle can be fitted back into its frame (the dots in the diagram show the effect of each rotation).

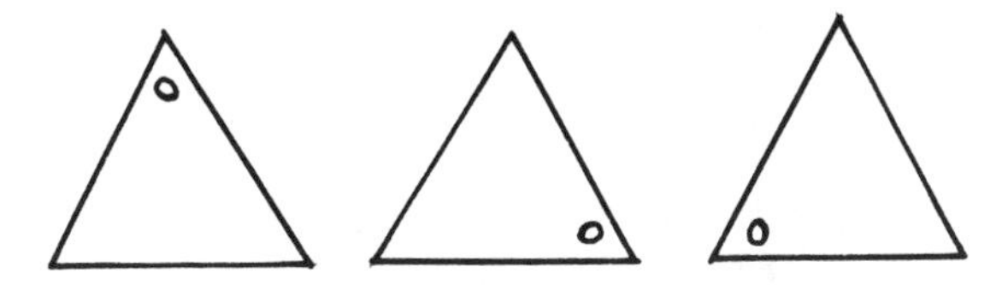

Since any shape can be fitted into its own outline in at least one way, then, mathematically, every shape has rotational symmetry of order 1. The yin-yang symbol with its traditional black and white colouring only has rotational symmetry of order 1: rotating it through 180° reverses the colours. But in outline only it has rotational symmetry of order 2.

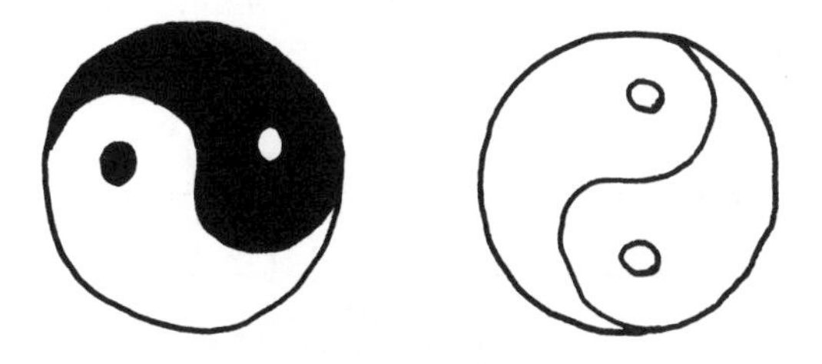

Transforming Shapes on a Grid

It's much easier to work out transformations to shapes if they are drawn on a grid. That way, you can work out the precise co-ordinates of where each line is in the original shape and its position in the transformed shape.

So, for example, here's a triangle *translated* four squares along and three up:

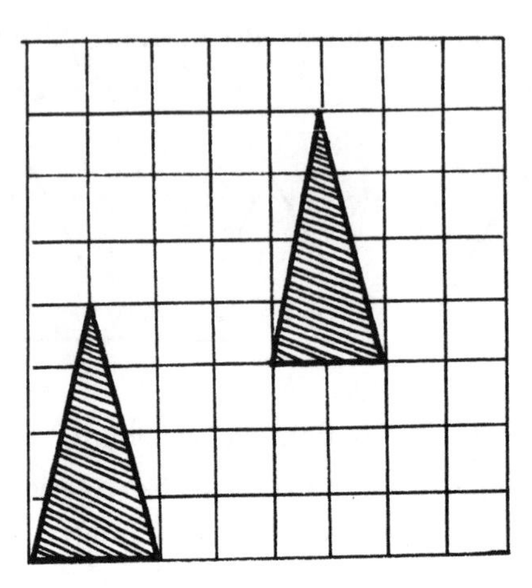

TEST YOURSELF

Here is that same triangle rotated 90° about . . . which point? Is it A, B or C?

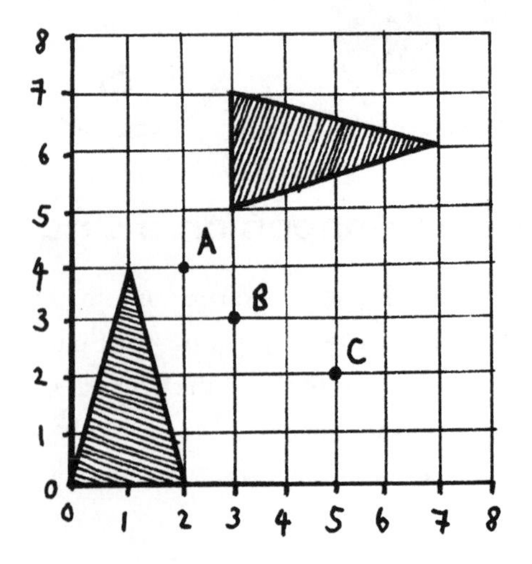

Vectors

The geometry of transformations can all be described very concisely using the language of co-ordinates and of *vectors.*

Vectors first began to enter the school syllabus in the 1960s (part of the 'New Maths') and although they now feature less prominently than they did, they are a part of maths that will be extremely important for any teenager who goes on to read engineering or to do technical design.

Vectors themselves are simply an arrow that tells you in what direction to go, and how fast or how far to go in that direction.

For example, on the previous page, we slid (or translated) a triangle four squares along and three up. This translation can be written in shorthand using a vector:

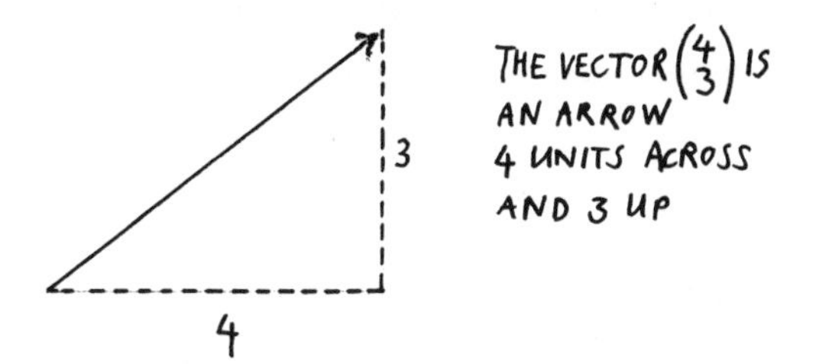

You'll notice this is just like giving co-ordinates, except that the numbers have been written vertically instead of horizontally. This is the mathematician's way of making sure that you don't get vectors and co-ordinates mixed up.

If you see two or more numbers in a column surrounded by brackets, they represent a vector.

Up to GCSE, vectors only ever refer to shapes in two dimensions, so they can always be represented on graph paper. They might be used to describe where a shape moves to, and they can also be used to describe a shape's speed. A balloon travelling horizontally at 6 metres per second is described by the vector:

$\begin{pmatrix} 6 \\ 0 \end{pmatrix}$ (A in the diagram below)

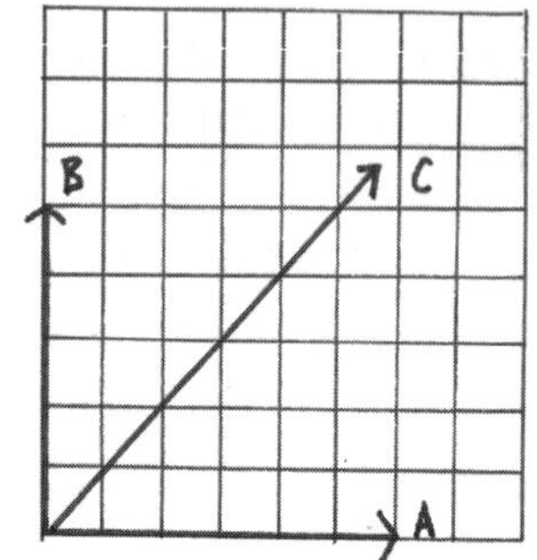

A balloon travelling at 5 metres per second vertically has the vector:

$\begin{pmatrix} 5 \\ 0 \end{pmatrix}$ (B in the diagram)

And a balloon travelling at 45° at 8 metres per second would have vector C in the diagram. Its length is 8 units (you can check), but the vector would be written as roughly 5.7 along and 5.7 up. The vector is actually the long side (hypotenuse) of a right-angled triangle, and to work out the lengths of its sides (the 'horizontal and vertical components') you'd need to use Pythagoras' theorem (page 167).

Adding vectors

Vectors can be added together. Doing this could not be simpler: just add the horizontal components and the vertical components together:

$$\begin{pmatrix} 5 \\ 2 \end{pmatrix} + \begin{pmatrix} 8 \\ 7 \end{pmatrix} = \begin{pmatrix} 13 \\ 9 \end{pmatrix}$$

More important perhaps is: why would anyone *want* to add two vectors together? As it happens, anyone involved in analysing moving objects (astronomers, air traffic controllers, engineers) will find practical uses for adding vectors together. Think of somebody travelling on a train (going at, say, 50mph) who opens the window and throws an apple core out (at 20mph).

Which direction does the apple actually travel over the ground? The train's forward motion is represented by the vertical arrow (50 units long) and the apple is thrown out of the train sideways (its arrow is 20 units long). The apple's actual path (the dotted arrow) is found by adding these two vectors together – joining them end to end. Using Pythagoras – or by measuring with a ruler – it turns out that the apple actually travels at about 54mph.

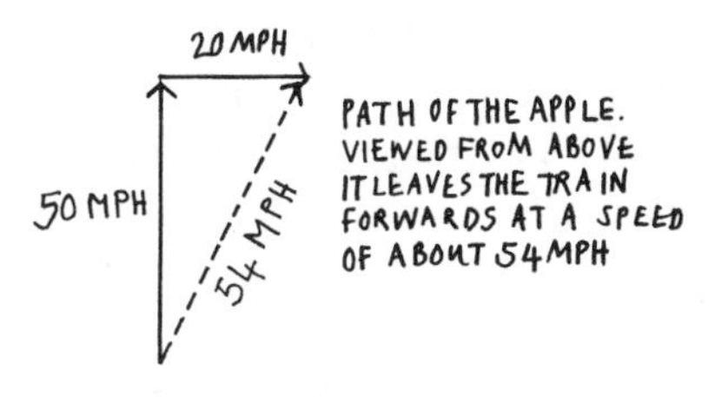

Something similar happens in rugby when a player passes the ball. If he is running forward and throws the ball sideways (relative to him), then the ball will actually travel forward – and if the ref spots it, he will blow the whistle for an illegal forward pass.

Who uses vectors?

2D vectors are really the path to understanding 3D vectors – across, up and out of the page towards your nose – and since we live in a 3D world, it's 3D vectors that have the most serious applications. Any profession involved in analysing 3D objects makes use of vectors; engineers and astronomers, for example. The film and games industry use them too: to look realistic, the objects in films and games have to mimic the way that light and shade looks in real life. Objects where the light is reflected straight back at you look brighter than those where the light is deflected off to the side. The only way for a computer to work out the direction that the light travels is by describing everything using vectors.

Computer graphics in school

Your teenager is most likely to explore many of the ideas in this chapter through 'dynamic geometry' computer packages. These are a bit like computer-assisted design packages but simpler, and they make it possible to explore geometry in a moving form, rather than based on static images.

The most popular dynamic geometry packages used in schools are called CabriGeometry and GeoGebra. Cabri is a commercial product, but GeoGebra is free to download and use at home.

Common Teenager Errors:

1) Difficulty in visualising the effect of mirror reflection on a shape. A particularly common problem is doing reflections in mirrors at an angle: they can find it hard to see that if a mirror line is at 45°, a horizontal line will reflect to become a vertical line.

2) When rotating a shape, getting confused between the centre of the shape and the *centre of rotation* of the shape. The centre of rotation can be anywhere, even outside the shape.

3) Working out where the centre of rotation is.

4) Thinking that parallelograms have line symmetry because cutting them along a diagonal produces two congruent (matching) triangles.

5) Misunderstanding the rules of scaling up a shape *(and this the is biggest, most common and most fateful)*, especially if the ratios are tricky. For example, if a 3 × 5 rectangle is being scaled up to 8 ×?, they will say 8 × 10; i.e. they have added 5 to both the width and the height. (The correct answer is 8 × 13.33)

MOVING AND CHANGING SHAPES
If you do only three things . . .

- Download GeoGebra and get your teenager, if they've worked with it, to teach you how to use it, or figure out together how it works.
- Encourage your teenager to use the shape-drawing facilities in word-processing programs to explore transformations. (All the regular programs like Microsoft Word allow you to translate, reflect and rotate objects.) Maybe you can persuade them to produce a design for you with a geometrical theme (the family Christmas card or a poster that you need, for example).
- When working on homework problems involving transformations, encourage your teenager to put the paper to one side and mentally visualise the figures in a question and to move them around in their heads. Going back to what is on the paper should gradually become clearer.

NUMBERS, CALCULATION & MEASUREMENT

There are three types of person:
those who can count and those
who cant.

Much of secondary-school maths provokes the petulant question 'When am I ever going to need this?', but not the content in this section. The relevance of numbers, calculation and measurement is obvious to anyone, though as it happens number theory – the investigation of some of the deep patterns that lie within the number system – is some of the most abstract 'maths for the sake of maths' that you could find. For example, the proof that there are different levels of infinity is just one of the mind-blowing concepts that emerge when you pursue maths beyond the confines of GCSE.

Tempting though it is for us to divert on to some of the big number ideas of maths, we've kept this section very practical. To be honest, for most teenagers there are more than enough mind-blowing ideas in understanding negative numbers, fractions and the properties of circles to keep them occupied.

At the end of this section, we've included a short chapter on calculators. It might be short, but it is extremely important, because most of the maths that teenagers will use beyond school is likely to involve the use of calculators (or their equivalents within computer spreadsheets and so on).

EXTENDING THE NUMBER LINE – NEGATIVES AND BIG NUMBERS

In primary school, most of the arithmetic that your teenager encountered involved the counting numbers, mainly in the range one to a thousand, but sometimes nudging into the millions. They probably met negative numbers and worked with fractions, but it isn't until secondary school that they do calculations with these and other numbers in earnest.

In this chapter we'll concentrate on the numbers that are outside the range one to a million: the negative numbers and very big numbers.

Negative Numbers

Every child these days learns to count with a number line. It starts from zero and counts horizontally to the right.

It's a very effective visual aid for counting forwards and backwards, and in secondary school it's a natural next step to extend that line to the left, introducing the negative numbers.

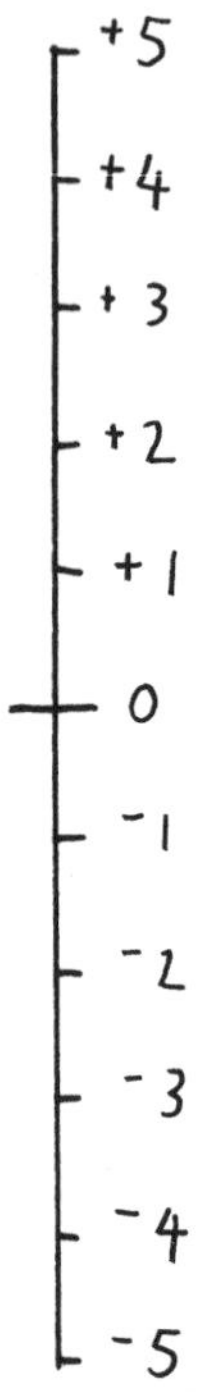

A VERTICAL NUMBER LINE IS A USEFUL WAY TO PICTURE NEGATIVE NUMBERS

Negative numbers have been used since the time of the Romans, though the Romans' own number system didn't allow for them, and for centuries afterwards most societies frowned on them. What began to establish negative numbers as useful was the idea that they can be used to represent debt. If you have £100 in the bank then they owe you that much money. If your account contains -£100, that is what you owe the bank, otherwise known as an overdraft.

Negative numbers are also very practical for indicating cold temperatures and for underground floors of a building. In fact visually it can be more helpful to picture the number line as being vertical, so that negative numbers become 'below zero'. That, after all, is how we describe freezing temperatures.

Together with zero, the positive and negative numbers produce a set of numbers that are known as the integers.

Arithmetic with negative numbers

So far, so good. But then your teenager is introduced to doing arithmetic with negative numbers and a whole world of potential confusion opens up. Why do 'two negatives make a positive' when you multiply? Why does taking away a negative number end up being the same as adding a positive? The short answer is 'because if you do that, everything works'. But that's not a convincing argument for a newcomer. Instead, it's good to build up an understanding in stages.

It can help to think of everyday language. The idea that two negatives make a positive is one we're all comfortable with. Think of: 'The biscuit is eaten' as positive, which makes 'The biscuit is *not* eaten' as its negative. If you now use *a double* negative to say: 'The biscuit is *not uneaten*', those two negatives cancel out, and the result is positive. 'The biscuit is not uneaten' is just a convoluted way of saying 'The biscuit is eaten'. Maths follows a similar logic.

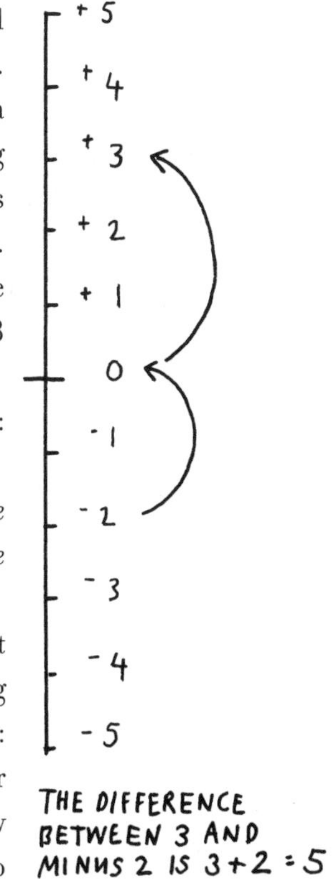

Negative numbers are represented with a symbol, −23, −6.1 and so on. Usually positive numbers aren't given a symbol, but in the early stages of learning the rules of positive and negative it helps to use a '+' symbol for positive numbers.

So, for example, when adding 3 + 2 we could write: $^{+}3 + {}^{+}2 = {}^{+}5$ (positive 3 plus positive 2 equals positive 5.)

In the same way, subtraction becomes: $^{+}3 - {}^{+}2 = {}^{+}1$ (in shorthand $3 - 2 = 1$)

What about subtracting a *negative* number: for example, 3 minus *negative* 2? We'd write this as: $^{+}3 - {}^{-}2 = ?$

It helps to think about what subtraction means. One meaning (which turns out to be helpful here) is: 'What is the difference between . . .'. Or in the language of travelling in a lift, how many floors would you need to travel to get from the floor that is underground 2 up to the third floor above ground? Two floors gets you to ground level then another three floors gets you to the third floor, making 5 altogether. This is one way of demonstrating why subtracting the negative of a number is the same as adding that number.

There are other ways of coming to this conclusion, for example by spotting the patterns and asking 'what comes next?' when you look at this sequence:

$$^{+}3 - {}^{+}3 = 0$$
$$^{+}3 - {}^{+}2 = 1$$
$$^{+}3 - {}^{+}1 = 2$$
$$^{+}3 - {}^{+}0 = 3$$
$$^{+}3 - {}^{-}1 = ?$$

It makes sense that the next number in the sequence will be 4, leading again to subtracting a negative number having the same result as adding the positive of that number. And if you continue with all the possible combinations you get this pattern:

Add a positive	$+ + = +$	$5 + {}^{+}4 = 9$
Add a negative	$+ - = -$	$5 + {}^{-}4 = 1$
Subtract a positive	$- + = -$	$5 - {}^{+}4 = 1$
Subtract a negative	$- - = +$	$5 - {}^{-}4 = 9$

Taking away negatives makes life more positive . . .

Some people like using an analogy between negative numbers and life. In life, positive things are good and negative things are bad. If you take away a good (positive) thing from your teenager's life, their phone, say, then their life gets worse (more negative). But if you take away a bad, or negative, thing from their life (those early-morning bangs on the door to get them out of bed) then they'll reckon that things get better. So it makes sense that subtracting a negative makes the result more positive. But let's not take this analogy too far . . .

Multiplying negative numbers

The rules for multiplying positive and negative numbers follow a similar logic. You can read 3×5 as 'three multiplied by five', in other words, $3 + 3 + 3 + 3 + 3$.

And so ${}^{-}3 \times {}^{+}5 = {}^{-}3 + {}^{-}3 + {}^{-}3 + {}^{-}3 + {}^{-}3 = {}^{-}15$.

So it's the same pattern emerging:

Multiply two positives	$+ + = +$	${}^{+}5 \times {}^{+}4 = {}^{+}20$
Multiply positive by negative	$+ - = -$	${}^{+}5 \times {}^{-}4 = {}^{-}20$
Multiply negative by positive	$- + = -$	${}^{-}5 \times {}^{+}4 = {}^{-}20$
Multiply two negatives	$- - = +$	${}^{-}5 \times {}^{-}4 = {}^{+}20$

It's still important to keep a clear head. What does it mean when you add two negative numbers together? For example, what is ${}^{-}3 + {}^{-}7$? When in doubt, turn it into real-world language: if you have an overdraft of £3, and add to it an overdraft of £7 what overdraft to you end up with? Adding an overdraft means you get further into debt, you become more negative, and the answer is $^{-}$£10. The first number in the calculation, $^{-}3$ in this case, is the position where you start on the number line, and the second tells you which direction to move along the line.

TEST YOURSELF

a) $^{-}7 - {}^{+}12 =$
b) $^{+}4 - {}^{-}12 =$
c) $^{-}3 \times {}^{-}7 =$
d) $3 - 2(5 - a)$

How negative numbers scuppered a Lottery scratchcard

In 2007, the National Lottery company Camelot introduced a winter scratchcard called 'Cool Cash'. The idea was that your card came with a freezing temperature on it (for example -7°) and you then had to scratch three windows to reveal three other temperatures. If any of the revealed temperatures was lower than the start number, you won the prize. The game hit the headlines when a woman called Tina from Lancashire bought a card where the target temperature was -7° and she scratched a window to reveal -6°. 'I've won because 6 is lower than 7,' she argued. Camelot disagreed, but decided to scrap the contest because of others having the same misunderstanding. Tina became a bit of a laughing stock because in the language of temperature, we understand 'lower' to mean 'colder' and -6° is slightly warmer than -7°. However, Tina did have a point. Very often when referring to numbers we use 'lower' to mean 'smaller'. Which is the smaller overdraft, £6 or £7? Of course it is £6, so there are times when $^{-}6$ is indeed 'lower' than $^{-}7$. The language of negative numbers has to be treated very carefully at times.

Very Large Numbers

Millions, billions and trillions are numbers that fill the news. What tends to get ignored is that there is a huge difference between them. If the width of your bedroom represents the number one billion, then one million is just the thickness of the plaster on the wall.

Your teenager is likely to meet numbers on this scale in subjects like geography, physics and economics, and when dealing with numbers of this size it can be hard to tell by inspection what the numbers are. With small numbers, like 346 and 2101, you can tell just by looking which is the larger or smaller, but with large numbers you need to do some counting of digits to know its size. For example, which of these numbers is the larger: 346392645221 or 75528965223? The answer is that the first is larger, but it probably took several moments for you to confirm it. Miscounting of digits and mis-entering a large number onto a calculator are two of the more common mathematical mistakes that teenagers make.

Where has the comma gone?

With larger whole numbers it is conventional now to put in spaces every three digits from the right, which makes them slightly easier to read: 346 392 645 221 or 75 528 965 223. It used to be the convention to put commas where the spaces are, but as countries like France use the comma to indicate the decimal point, commas are now out of favour when writing out large numbers and you probably won't see them in your teenager's textbook.

However, spaces are also open to ambiguity. Is 346 892 a number with hundreds of thousands, or is it two separate numbers, 346 and 892? To add to the confusion, the convention for numbers after the decimal place is simply to write out the digits with no gaps or commas.

Powers

On page 131 we introduced the idea of exponential numbers. For example, 2^3 means $2 \times 2 \times 2 = 8$. Exponentials can be used to express incredibly large numbers. 2^{270} might look innocuous, but it actually represents an astronomical number – literally so, as it is roughly the number of atoms in the entire universe.

Powers aren't just a useful shorthand for large numbers, they can also simplify calculations.

To work out, say, $3^4 \times 3^2$, all you need to do is add the small numbers (indices) to get the answer: 3^6. Written out in longhand it's easy to see why this works: $(3 \times 3 \times 3 \times 3) \times (3 \times 3) = 3 \times 3 \times 3 \times 3 \times 3 \times 3$, you are simply counting how many times the number 3 has been multiplied together.

Using similar logic, you can divide powers simply by

subtracting the indices. $7^8 \div 7^3 = 7^5$. This rather neat trick of turning multiplication into addition, and turning division into subtraction, is also known by the rather more scary name of *logarithms* (a word rarely encountered until A level these days). It works for any number you like, including fractions, so for example $\frac{1}{2}^3 \times \frac{1}{2}^4 = \frac{1}{2}^7$, and in general $a^x \times a^y = a^{(x+y)}$. Of course, it looks more intimidating to teenagers when they see it written in algebraic form, and as with many examples in maths, it can be helpful to substitute in numbers when wanting to check what the rule is.

It's an easy and common mistake to start wanting to use this to do calculations where the base numbers are different, for example 3 and 5. What is: $3^2 \times 5^3$? A teenager might want to write the answer 15^5 but the simple rule of adding the indices only works if the same number 'a' is being raised to a power in the multiplication or division.

The simplest and most useful powers are the powers of 10. For example, $10^2 = 100$ and $10^4 = 10\,000$.

It's a nice feature of the powers of 10 that you can work out the power simply by counting the number of zeroes. What's 10^9? It's a one followed by nine zeroes, 1 000 000 000, otherwise known as a billion. This even works for zero: 10^0 is 1 with no zeroes, in other words '1'!

The power of zero

What is 10^0? Most teenagers would naturally expect the answer to be zero, because it seems to be saying '10 multiplied by itself zero times'. Mathematicians could indeed have decided to say that 10^0 equals zero (in the abstract world of maths you can in theory define symbols in any way you like), but in fact it fits far better with the rest of maths if you instead decide that all numbers raised to the power zero equal 1. Why? Because of the rule about adding indices. If $10^4 \times 10^5 = 10^9$, then what is $10^4 \times 10^0$? By the adding rule, since $4 + 0 = 4$, the answer must be $10^4 \times 10^0 = 10^4$, in other words 10^0 must equal 1. This is true for all numbers. $3^0 = 1$ and $\frac{1}{2}^0 = 1$. Once you accept this, you can start to explore the meaning of negative powers such as 10^{-2} and discover that even these make sense. More of that on page 221.

There's one question that troubles many older teenagers. They know that zero to the power of anything is always zero, yet they are told that all numbers raised to the power of zero equal one. So what is 0^0, is it zero or one? The answer (mathematicians will tell you) is that either it is meaningless, or that 'it depends' – an illustration that maths does not always have a definitive answer.

Significant figures

When Tesco announces its annual turnover, the press could report it at (say) 65 931 242 103 pounds and 73 pence. But they don't, because obviously we don't need to know the figure in that much detail. Indeed, only a few people (accountants and physicists come to mind) ever need to know more than the first few digits of any number. Big numbers are easier to cope with when they are rounded to a few 'significant figures'. That fictitious Tesco number above would probably be rounded to the value £65.9 billion, which is three significant figures followed by zeroes.

The most significant digit in any number is the first digit in the number that isn't zero, since that digit represents the largest part of the number. For example, in the number 3846, the most significant figure is the 3 which represents 3000. In the number 26, the most significant figure is the 2, while in 0.0582, it is the 5 (which represents five hundredths).

Rounding a number to significant figures means rounding it to the nearest value, and there's a slight catch here that confuses teenagers. In order to round a number to two significant figures, you have to look at the first *three* digits. To round it to three significant figures you have to look at the first *four* digits. In other words, you always have to look at one more digit than the number of significant figures you are rounding to.

To round 3846 to two significant figures means looking at the first three digits 384. If the last digit is less than 5 (as it is in the case of 384) you round that digit down to zero, making 3800. However, had the third digit been 5 or more, you would round the number up to 3900.

So 3846 to two significant figures becomes 3800 (the 4 is rounded down) whereas to three significant figures it becomes 3850, as the 6 needs to be rounded up.

By the way, once you have found the first significant digit in a number, all the other digits that follow in the number are significant, even if they are zero. In the number 4008, both zeroes are significant, so when rounded to two significant figures this number is 4000, while to three significant figures it is 4010. The idea that zeroes are sometimes significant and sometimes not (when they start a decimal, for example) is a regular source of confusion for teenagers.

Standard form

To reduce the chance of making errors, and also to save time when writing and manipulating large numbers, mathematicians and scientists much prefer to use a number shorthand known as *standard form*.

Take the number 384 703, for example. The first '3' here represents 300 000, which can also be written as $3 \times 100\ 000$. The whole number can be written as $3.84703 \times 100\ 000$. More conveniently, 100 000 can be written as a power of 10, in this case 10^5.

So the number 384 703 can be written as 3.84703×10^5 , and this is what is known as standard form. It is 'standard' in the sense that this is the standard notation that scientists across the world like to use for writing numbers. You could, incidentally, also write the number in countless other ways, for example 38.4703×10^4, but it is only standard form when written with one digit before the decimal point.

This notation starts to become useful with very large numbers, such as:

$341\ 392\ 645\ 221 = 3.41392645221 \times 10^{11}$

while $75\ 578\ 965\ 223 = 7.5578965223 \times 10^{10}$

Without any counting, this standard form of expressing the numbers tells us immediately that of the two numbers above, the first one is the larger one.

More commonly, these numbers will only be quoted to two or three significant figures, for example:

3.41×10^{11}

and 7.56×10^{10}

Multiplying and dividing using standard form

Standard form comes into its own when doing calculations. To calculate 340×2000, you can first convert it to standard form: $3.4 \times 10^2 \times 2 \times 10^3$ and then multiply the small numbers together before adding the powers of ten: 6.8×10^5.

Dividing is equally simple:

$$8400 \div 210 = \frac{8.4 \times 10^3}{2.1 \times 10^2} = 4 \times 10^1 = 40$$

TEST YOURSELF

a) Write the number 24 892 in standard form, to three significant figures.

b) Without a calculator, work out $2.3 \times 10^4 \times 2.0 \times 10^5$ (and give the answer in standard form).

c) What is $(9 \times 10^9) \div (3 \times 10^7)$?

Who wants to be a million area?

Several years ago, a 'big number' question cropped up on the popular TV show *Who Wants to be a Millionaire?* The question posed was this:

WHICH OCEAN HAS AN AREA OF 5 MILLION SQUARE MILES?

A) ARCTIC B) INDIAN

C) ATLANTIC D) PACIFIC

The couple taking part, already sitting on several thousand pounds in possible prize money, decided to ask the audience. Perhaps not surprisingly, the audience's opinions were split. About half plumped for the Pacific, followed by the Atlantic, the Indian and the Arctic. This was not the definitive audience response that the contestants were hoping for, and they sensibly chose to take the money and leave.

Almost certainly the thinking of most members of that audience went like this: 'Five Million is really huge. The Pacific Ocean is really huge. So it's probably the Pacific.' A back-of-envelope estimate shows why they were wrong. Let's think about the ocean that the British are most familiar with, the Atlantic. Very crudely you can think of the Atlantic as a rectangle of water running between the poles and between the Americas and Europe/Africa. How tall is this rectangle? The circumference of the world is about 24 000 miles, so 10 000 seems a reasonable estimate for the north-south height of the Atlantic. And let's say it's about 3000 miles across (it takes about five hours to fly from London to New York, and planes travel at about 600 mph). That would make the area of the Atlantic 10 000 × 3000. Using standard form it works out at 3×10^7 or 30 million square miles – far bigger than the 5 million mentioned in the question. The Pacific is even larger. In fact, it turns out that for an ocean, 5 million square miles is rather small, and the only ocean small enough to be that size is the Arctic. It's a powerful reminder that some big numbers are really quite small.

EXTENDING THE NUMBER LINE

If you do only three things . . .

- Help your teenager remember how to multiply or divide with positive and negative numbers by remembering that 'like' signs always combine to be positive. If the signs of the number don't match, multiplying or dividing gives a negative answer.
- Encourage them to practise the Zequals technique of estimation (page 41) using standard form. Remind them that Zequals only uses numbers to one significant figure.
- Look out for large numbers being quoted in the news and work with your teenager in expressing these in different ways.

DECIMALS, FRACTIONS AND OTHER SMALL NUMBERS

There are four common ways that numbers between zero and one are presented. There's the simple fraction, such as $\frac{3}{8}$; the decimal fraction, for example 0.375; there's the percentage, 37.5% (which is $37.5 \div 100$); and then there's standard form, such as 3.75×10^{-1}.

As you might have spotted, the four examples we just used all represent the same number. So why all the choice? Partly it's down to personal preference – there is no law that requires one form be used ahead of the other – but it's also a matter of horses for courses. There are times when it's easiest to work with fractions, particularly if you are multiplying. Tiny numbers are often dealt with most easily using standard form – the wavelength of a deadly gamma ray is 0.00000000001 metres, much easier written as 1×10^{-11}. Percentages are ideal when comparing fractions with each other.

Teenagers need to be able to comfortably handle all of them, and also to be able to convert from one form to another.

Decimals and Percentages

Decimal places in a number are first covered in primary school, but they continue to cause problems to pupils all the way through secondary school. One common problem is forgetting what the different positions in a decimal fraction represent, so here is a reminder, using the decimal 0.1768 as an example.

In the same way as the digits of a whole number to the left of the units represent the tens, hundreds, thousands, and so on, by nice symmetry the numbers to the right of the units, after the decimal point, represent tenths, hundredths, thousandths, etc.

Writing the number 0.1768 into columns we get:

UNITS		TENTHS	HUNDREDTHS	THOUSANDTHS	TENTHOUSANDTHS
0	.	1	7	6	8

We could also write the number as '1.768 tenths' or '17.68 hundredths' or '176.8 thousandths' and so on, all of them fractions: 1.768/10 = 17.68/100 = 176.8/1000 = 1768/10000. As it happens, one of these is a fraction we use all the time: 17.68/100 is 17.68 'per cent'.

Confidence in being able to read off decimals as percentages (and the reverse) saves your teenager ever needing to use the dreaded percentage key on a calculator, a button that causes far more problems than it is worth.

TEST YOURSELF

a) Which is the larger number, 0.342 or 0.3419?
b) What is 0.042 as a percentage? And what is 1.07 as a percentage?

Converting Between Fractions, Decimals and Percentages

Many teenagers don't pick up on the fact that a fraction is simply a division: $\frac{3}{8}$ means 3 divided by 8, and they need reminding that 'per cent' is also a fraction, it means 'divided by 100'. When converting the fraction $\frac{3}{8}$ into a percentage it's then a matter of realising that: $3 \div 8 = ? \div 100$ where '?' represents the percentage.

Being able to convert a ratio in this way is one of the essential everyday maths skills that we discussed on page 48. In this case, we can work out what '?' is by multiplying $3 \div 8$ by 100, the answer being 37.5.

But how do you actually divide 3 by 8 to produce the number 37.5? By default, most people would leap straight to a calculator, but simple percentages can easily be worked out mentally or on a piece of paper. Your teenager will learn to do this in various stages.

a) Simple fractions: $\frac{1}{2}$ and $\frac{1}{4}$ are easily converted into percentages as the conversion is straightforward. $\frac{1}{2} = 50/100 = 50\% = 0.5$ while $\frac{1}{4} = 25 \div 100 = 25\% = 0.25$. What makes these easy is that the bottom number on the fraction (the denominator) divides into 100 exactly. Fifths, tenths, twentieths and so on are also therefore easy to turn into percentages (for example, $3 \div 20 = 15 \div 100$, or 15%)
b) It is not so obvious how to convert a fraction like $14 \div 35$ into a percentage – until you realise that the fraction can be simplified. In this case, divide the top and bottom by 7, and you get $2 \div 5$, which is $40 \div 100$, or 40%.
c) Then come the more challenging fractions, such as $1 \div 9$

or 7 ÷ 26. To convert these into decimals and percentages, what's needed is some old-fashioned division.

Short division

Although 'short division' and 'long division' (see page 220) have never gone away, for many years they have been rather out of fashion. In primary schools, children are introduced to division using a technique known as chunking. Chunking is a useful way of building an understanding of division, and is a helpful tool for estimating answers. It treats division as being like a repeated subtraction. For example, if you want to calculate 182 ÷ 7, you work out how many times you can subtract 7 from 182. You could literally subtract 7 twenty-six times but that would be incredibly tedious, it is much faster to remove large chunks of 7, for example in chunks of 70. Here's how one child worked it out:

```
7   182
   - 70    10
   ----
    112
   - 70    10
   ----
     42
   - 42     6
   ----
      0              10 + 10 + 6 = 26
```

With more confidence, a child will spot that you can remove a chunk of 140 (that's twenty 7s), and then a chunk of 42 (six 7s), which shortens the calculation.

By the time they reach secondary school, most children have been introduced to short division, more commonly known these days as the *bus-stop method* (because the division box slightly resembles a bus shelter).

To calculate $182 \div 7$:

$$\begin{array}{r} 2\;6 \\ 7\,\overline{)\,1\;{}^{1}8\;{}^{4}2} \end{array}$$

There is a standard mental 'script' that adults traditionally use, which in this example would go: '7 into 1 doesn't go, carry 1, 7 into 18 goes 2 (write 2 above the line) carry 4, 7 into 42 goes 6 (write 6 above the line).

It may look different from chunking, but it is just a condensed version of the same thing. First you remove as many chunks of 70 as you can (20 goes on the top line) then you remove as many chunks of 7 from what is left, $182 - 140 = 42$ (6 goes on the top line), to give the answer 26.

We pointed out in *Maths for Mums & Dads* that the disadvantage of this method is that for most children it is merely a black box that works for reasons they don't understand. As a result, many quickly forget how to do it, and without regular practice they only half remember the rules, and so tend to make lots of mistakes.

However, for those many teenagers who do 'get it' and learn it, the bus-stop method has some big advantages too. It is concise. It is quick. It keeps all the numbers in the right 'place value' column (one column represents hundreds, the next is tens, then units, and so on). And when it comes to calculating decimals by hand (or mentally), the bus-stop method is without doubt the most efficient method available. Also, if you do learn it and get confident at it (and many teenagers do), it will save needless use of a calculator and will also build greater confidence in handling fractions, decimals and percentages.

Turning fractions into decimals

If you want to turn a fraction such as $\frac{3}{8}$ into decimals without a calculator, the best method is short division. The only difference from divisions into larger numbers is that you need to write in zeroes after the decimal point, in order to keep track of decimal places.

To calculate $\frac{3}{8}$, write 3 as 3.000 inside the bus stop (three zeroes after the decimal point is enough to be starting with for any decimal calculation).

$$\begin{array}{r} 0\cdot3\;\;7\;\;5 \\ 8\,\overline{)\,3\cdot{}^{3}0\;{}^{6}0\;{}^{4}0} \end{array}$$

The 'script' goes: '8 into 3 doesn't go (write 0 followed by a decimal point), carry the 3; 8 into 30 goes 3 times (write 3 next to the first zero), remainder 6 (write 6 by the next zero); 8 into 60 goes 7 times (write 7 above the 60), remainder 4, 8 into 40 goes 5 (write 5 above the 40). No more remainder so that's it.'

You can use this short, bus-stop division for any division at all. Here is how to work out 11 ÷ 24 as a decimal:

$$\begin{array}{r} 0\cdot4\;\;5\;\;8\;\;3\ldots \\ 24\,\overline{)\,1\;\;1\cdot{}^{11}0\;{}^{14}0\;{}^{20}0\;{}^{8}0\ldots} \end{array}$$

There are two things to notice. First, we haven't finished the calculation. In fact, written as a decimal, 11 ÷ 24 never ends, so we have to round it, typically to three decimal places, so it is 'about 0.458'.

The second thing is that because the divisor (the bottom number of the fraction) is bigger than 10, the remainders carried over are bigger than 10. This means that the whole calculation gets messy and congested with numbers.

So if you are doing a precise calculation of a decimal with a large divisor by hand, it can be better to use the method known as long division.

Long division

There are those who say: 'Anyone who has ever done two long divisions in their life has done one too many.' Long division is time-consuming, and is only ever mastered by a small minority of school pupils. Furthermore, by the time you get beyond GCSE, life is too short to work out 23 ÷ 87 with paper and pencil. If you can do it in your head, as a few people can, or if you simply enjoy doing long division as a mental workout, then that's great, stick with it. Frankly, if a calculation is complicated enough to require long division then of course you should use a calculator.

Just for completeness, however, let's remind you of how 11 ÷ 24 would be written out as a long division. (The calculation 110 ÷ 24 would use exactly the same method, but with the decimal point one place to the right at every stage.)

```
         0·4 5 8 3...
   24)1  1·0 0 0 0...
         9·6                Write decimal point, then 4 × 24 = 96
        ─────
         1·4 0              110 − 96 = 14, bring down a zero to make 140
         1·2 0              5×24 = 120
        ───────
          ·2 0 0            140 − 120 = 20, bring down a zero to make 200
          ·1 9 2            8 × 24 = 192
        ─────────
          ·0 0 8 0          etc., etc.
          ·0 0 7 2
```

Decimals and Standard Form

One of the most common problems teenagers have with decimals is miscounting zeroes, just as with large numbers. The other common problem is forgetting what the different positions in a decimal fraction represent.

For example, what is 0.000007 expressed as a fraction? You probably find yourself counting from the left to realise that it represents 7 millionths, or 7÷1 000 000 (but it's easy to mistakenly get, say, 7 hundred-thousandths).

When dealing with decimals that have leading zeroes, mathematicians sensibly tend to adopt standard form (which we introduced on page 210). In fact, when it comes to doing multiplication or division with decimals, standard form is almost essential.

Remember that tens, hundreds, etc. can be written as 10^1, 10^2, and so on. In the same way, the decimal places can be represented by the *negative* powers of 10: one tenth is 10^{-1}, one hundredth is 10^{-2}, and so on.

Teenagers typically expect a number such as 10^{-1} to represent a negative number, so what's it doing meaning 0.1? The answer as always with these indices is 'because it works', and you can test it using the rule of adding indices when you multiply. What is $10^2 \times 10^{-1}$ (i.e. 100×0.1)? By the adding rule, $10^2 \times 10^{-1} = 10^1$ (= 10). Does this make sense? Yes, because $100 \times 0.1 = 10$.

Writing the number 0.1768 into columns we get:

UNITS		TENTHS	HUNDREDTHS	THOUSANDTHS	TENTHOUSANDTHS
0	.	1	7	6	8

The most significant digit in 0.1768 is the '1', as it represents one tenth (or ten hundredths). Next comes the 7, 6, 8, and so on.

In standard form, one tenth is 1×10^{-1}; seven hundredths is 7×10^{-2}; six thousandths is 6×10^{-3}; and eight ten-thousandths is 8×10^{-4}.

The whole number can be written in standard form as 1.768×10^{-1} (or in non-standard form as 17.68×10^{-2}, or 0.01768×10^{1} or countless other ways).

Significant figures work just the same way as for whole numbers. The decimal fraction above is 0.18 to two significant figures (notice that the 7 has been rounded up because of the 6 thousandths), which is 1.8×10^{-1} in standard form.

It gets a bit harder when the decimal has lots of leading zeroes. Take for example: 0.000000004632.

This is $4.632 \times$ ... what? The easiest way to find out is to point your finger between 4 and 6 (as if it's the decimal point) and then count how many digits you have to move it to the left to get to the correct position for the decimal point. In this case you count nine to the left. So in standard form the number is: 4.632×10^{-9}.

Multiplying and Dividing by Decimals

Multiplying and dividing by decimals using standard form is exactly the same as for larger numbers, though with the complication that it requires adding and subtracting of negative numbers (explained in the previous chapter).

For example, 4×10^{5} multiplied by 2×10^{-3} means multiplying 4×2 (=8) and then adding the indices: $^{+}5 + {}^{-}3 = 2$ (because adding a negative is the same as subtracting). So the answer is 8×10^{2} (= 800). Written longhand, this calculation is 400 000 $\times$ 0.002 = 800 but setting it out as a long multiplication is asking for trouble – the chance of making an error is very high.

And what about a calculation like this? $0.6 \div 0.0003$? Of course any teenager would be tempted to leap for a calculator, but there are plenty of ways to do it without a calculator, for example using standard form:

$$\frac{6 \times 10^{-1}}{3 \times 10^{-4}}$$

$6 \div 3 = 2$; and to calculate $10^{-1} \div 10^{-4}$ means working out $^{-}1 - {}^{-}4$, which is the same as $^{-}1 + 4 = {}^{+}3$, so the answer is 2×10^3 (or 2000).

That's probably as hard as anything your teenager will encounter at GCSE. But you know what: it isn't *that* hard.

Common errors with standard form of decimals

1) Putting more than one digit in front of the decimal point – 0.083 should be written as 8.3×10^{-2} not as 83×10^{-3}.
2) Miscounting the number of places to shift the decimal point, especially with lots of leading zeroes. The mistake is often to count the number of zeroes rather than the number of positions that the decimal point has to move. For example, the number 0.0007 will often be wrongly recorded as 7×10^{-3} (it is actually 7×10^{-4}).
3) Thinking that 'zero' is never a 'significant figure'. (Zero is never the FIRST significant figure but it can be significant after that.) 0.6072 is 0.607 to three significant figures and 0.030004 is 0.0300 to three significant figures.
4) Forgetting the rules for adding and subtracting negative numbers.

TEST YOURSELF

Express each of these numbers in standard form to three significant figures.

a) 0.0309374302
b) 0.0008117

Multiplying Fractions

Just as adding and subtracting negative numbers causes grief for many teenagers so too does multiplying and dividing by fractions. The grid method for multiplication (see page 87) is a useful way of explaining how multiplication of fractions works. Suppose we want to multiply $\frac{5}{8}$ by $\frac{4}{7}$. Draw the grid dividing along the top into eight sections (representing the eighths) and the side into seven sections (representing sevenths).

To do the multiplication we need a bit of the grid that is five of the eighths along the top, and four of the sevenths down the side.

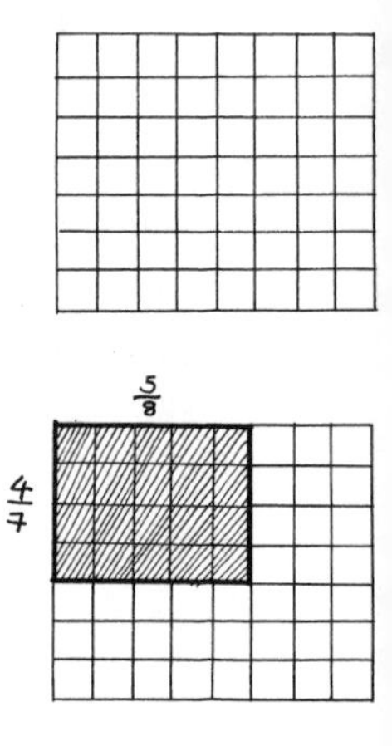

This grid has been divided into 8 × 7 (= 56) pieces, of which 5 × 4 (= 20) have been shaded. So the new fraction is 5 × 4 (= 20) out of 8 × 7 (= 56) or $\frac{20}{56}$. This method works for any pair of fractions and is a visual way to see why the rule for multiplying fractions is to multiply the top numbers (the numerators) and multiply the bottom numbers (the denominators).

$$\frac{5}{8} \times \frac{4}{7} = \frac{20}{56}$$

Dividing by Fractions

Dividing by fractions is more complicated. The rule that people are usually taught is 'turn upside down and multiply'. So for 3 divided by $\frac{5}{8}$ we would invert $\frac{5}{8}$ to make $\frac{8}{5}$ and then calculate: $3 \times \frac{8}{5}$.

Why does this work? Let's start with a simpler example, 3 divided by $\frac{1}{8}$. If we think of division as repeated subtraction – that is, how many eighths are there in 3? – then the answer is 24: there are eight eighths in 1, so 24 eighths make up 3. So dividing by a simple unit fraction (a fraction with one as the numerator) is the same as multiplying by the denominator.

$$3 \div \frac{1}{8} = 3 \times \frac{8}{1} = 24$$

Dividing by $\frac{5}{8}$ instead of $\frac{1}{8}$ means the answer will be five times smaller. In other words the top part of the fraction you are dividing by goes on the bottom when multiplying:

$$3 \div \frac{5}{8} = 3 \times \frac{8}{5} = \frac{24}{5}.$$

But while it's nice to know there is a reason why division by fractions works this way, most teenagers will prefer to move on and just learn the method. Perhaps the Americans got it right. They remember the rule as:

Ours is not to question why
Just invert and multiply.

TEST YOURSELF

Calculate:

a) $\frac{3}{4} \times \frac{4}{5}$

b) $\frac{2}{3}$ of $\frac{7}{9}$

c) $6 \div \frac{1}{3}$

d) $\frac{5}{6} \div \frac{5}{8}$

Endless Decimals and Irrational Numbers

We saw on page 215 that many fractions, such as $\frac{1}{10}$ or $\frac{3}{8}$, make nice neat decimals, 0.1 and 0.375 here. But that isn't always the case. Sometimes a simple fraction forms a decimal that goes on for ever. One example is $\frac{1}{9}$, which makes 0.1111111 . . . (Instead of a stream of dots, mathematicians like to place a single dot over the first recurring decimal, like this: $0.\dot{1}$, which means the 1s go on for ever.)

Teenagers are often intrigued by the idea that if $\frac{1}{9} = 0.11111$, and you then multiply $\frac{1}{9} \times 9$, the answer is 0.999999... . Yet surely $\frac{1}{9} \times 9 = 1$. In fact, mathematicians tend to argue that this demonstrates that 0.99999... is actually *the same* as 1.

Other decimal fractions can be more complicated. $\frac{1}{7}$ as a decimal is 0.1428571428571 . . . with the pattern 142857 repeating forever. The shorthand for indicating the repeating sequence is to place dots over the digit that starts and ends that sequence, so: $0.\dot{1}4285\dot{7}$

Any fraction that is a whole number divided by another whole number (known as *rational* numbers because they are a ratio of two numbers), be it $\frac{27}{92}$ or $\frac{844}{973}$, will form a decimal that either stops after a certain number of digits or has a pattern that repeats for ever. All of these rational numbers can be added on to the number line:

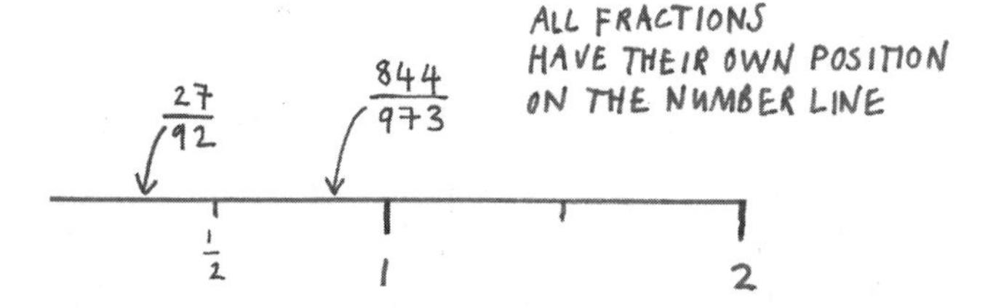

If you draw the number line larger, you can always fit more fractions in. In fact, given any two rational numbers another can always be squeezed between them. Take, for example, $\frac{1}{8}$ and $\frac{1}{7}$. Putting these over a common denominator they can be expressed as $\frac{7}{56}$ and $\frac{8}{56}$. We can then squeeze $\frac{7.5}{56}$ (or $\frac{15}{112}$) between these two. From this it begins to become clear that there is going to be a never-ending (or infinite) number of rational numbers occupying the number line because you can always add more. And yet however many rational numbers you add to the line, there will, incredibly, always be gaps, because there are infinite other numbers squeezed among them. These are the so-called *irrational* numbers, irrational not because they don't make sense but because they cannot be expressed as a simple ratio of two whole numbers.

One example of an irrational number is the square root of two, $\sqrt{2}$. This number is approximately 1.414. There are plenty of fractions that are very close to $\sqrt{2}$, for example $\frac{17}{12}$, but no simple fraction is *exactly* $\sqrt{2}$.

If this is beginning to baffle you, then you are not alone. Over the centuries many mathematicians have struggled to grasp the ideas behind the infinity of irrationals. It gives just a hint of the many deep mysteries that lie within the world of numbers.

DECIMALS AND FRACTIONS
If you do only three things . . .

- Get your teenager to commit to memory common 'equivalents' – half, $\frac{1}{2}$, 0.5, 50% are all the same thing. So are $\frac{1}{4}$, 0.25, 25% and 'a quarter'. With these as 'building blocks', many calculations become easy (25% of 448, that's a quarter: 112).
- Remind them that in mathematical calculations, 'of' almost always means 'multiplied by', so 'half of 12' means $\frac{1}{2} \times 12$, while '32 per cent of the population' means 'the population multiplied by $32 \div 100$'.
- Encourage your teenager to think of fractions as numbers in their own right – 'two thirds' isn't made up of a separate 2 and 3, it is a point on the number line that's two thirds of the way from zero to one.

MEASUREMENT

There are, it might be said, two types of thing in the world. There are things that can be counted (such as cows), and things that can be measured (such as milk, or the level of mooing).

Most people figure out counting when they're quite young, but measuring causes a great deal more grief, through teens and into adulthood. It's probably true that more mistakes are made in measurement than in any other aspect of everyday maths. Perhaps the most dramatic and costly of these was the Mars Orbiter disaster described on page 232.

Teenagers tend not to have much problem linking measurement maths to the real world, but that doesn't mean they don't fall into a wide variety of pitfalls when trying to answer problems. In primary school, children learn about the units of measurement for time, distance and weight separately. In secondary school, these measures start to be combined: speed is distance divided by time, density is mass divided by volume, and so on. These more complicated measurements can't just be read off with a ruler or a stopwatch, they need to be calculated, and that's when the errors begin to creep in.

Measurement doesn't just feature in maths lessons, it crops up in almost every other subject, including the sciences, geography, design technology and art. This can be part of the problem, since the geography or physics teacher might assume that 'logarithmic scales on a graph' (for example) will be taught in maths, while the maths teacher might reckon that it really belongs in physics – the result being that it falls between the cracks and never gets taught properly at all.

So be warned, unlike the other chapters in this book, questions about the topics in this chapter might arise from homework in a subject that isn't maths.

The Three Aspects of Measurement

There are three main aspects of measures that your teenager will meet.

- The common-or-garden measures that form the core of the metric system for measuring length, volume, weight and so on – the simple, standard units.
- Measuring units that arise from combining two standard units, for example, density – a combination of mass (weight) and volume – or speed (combining distance and time). These are known as the compound units, and as we will see below, they are a major source of confusion.
- Finally there is a collection of ideas around area, perimeter and volume, and within this comes the particular (some would say peculiar) case of the circle and its area and perimeter (aka circumference) and the mystical number pi that links these two things.

Simple Measures, Metric and Imperial

It's about fifty years since Britain made the first steps towards metrication, and school maths has been almost entirely based around the metric units – metres, litres and kilograms – since then. So you might assume that today's teenagers think in these units all the time. If so, you would be wrong. We gave the following quiz to fourteen-year-olds in a range of schools across the country:

1) How tall is George Clooney?
2) How far is it from London to Edinburgh?
3) How much does Kylie Minogue weigh?
4) How much beer is a 'safe' weekly intake?
5) What's the width of a lane on an Olympic athletics track?
6) How much water should you drink in a day?

Of course they weren't expected to *know* the answers to these questions and there was often quite animated discussion, though in fact most teenagers we asked came up with very reasonable estimates. What was striking, however, was the units that they used in their answers. You might like to have a guess before reading on.

Consistently across all the groups that we asked, imperial units were used more often than metric. George Clooney's height? 'About six feet.' London to Edinburgh? 'Five hundred miles.' Kylie Minogue? 'Eight stone.' Weekly beer consumption? 'Ten pints.' Only the last two questions produced metric answers. An Olympic lane width was typically reckoned to be 'one metre' and daily water consumption 'a litre'. The reason why they use imperial units for so many everyday measures is that their parents use them (and in the case of distance and beer, imperial is still the standard unit in official use).

Metric units have made mathematical calculations far easier than the imperial units that their grandparents used, but our survey shows that imperial units are still common currency across all age groups. And while teenagers have a reasonable feel for imperial units in isolation, when applied to things they are familiar with, their ability to convert from imperial to metric is quite shaky. When asked to convert their answers for beer and water consumption from pints to litres and vice versa, many teenagers struggled.

One teacher told us of the annual fluster that is created when her school goes on the annual trip to France: *Our Year Tens have to fill in forms giving their height in centimetres and weight in kilograms, and they are always asking for reminders as to how to do it.*

The Mars Orbiter – an expensive maths error

In September 1999, the Mars Orbiter was nearing its destination, its mission being to discover more about the climate on the red planet. Unfortunately there was a catastrophe – the Orbiter entered the wrong trajectory and crashed into the planet's surface. Investigators later discovered what had happened: two teams working on the specification for the spacecraft had used different units. One team was working in metric units, while the other was working in imperial units of feet and pounds (still common practice in the USA). When the computers calculated the thrust needed in the rockets to get the craft into orbit, they gave it a number that was 4.5 times too large, exactly the ratio of 'pound force' to Newtons, with disastrous consequences. The cost of the failed mission was more than $300 million, making this one of the most expensive maths errors of all time.

Measurements on a Human Scale

Our ancestors invented imperial units for a reason: they are designed to be numbers that work on a human scale. One foot is roughly the length of a man's foot. The height of a tall adult male is about six feet, and if he's overweight he's about 15 stone. If he's unwell his temperature might be 100°, he fills up his car with about ten gallons and when he goes to the pub he drinks a pint. Their rounded metric equivalents – 30cm, 180cm and 90 kilograms, a temperature of 37°C, 50 litres of petrol, and 600ml of beer – are usually more unwieldy numbers. And as an aside, metric units are less lyrical too. Dozens of songs refer to miles, feet and gallons but you'll struggle to find any song that mentions kilometres or litres.*

Converting from one unit of measurement to another is one place where an understanding of ratios is important. Sometimes a precise conversion is needed, but just as often a rough-and-ready conversion is good enough. Fortunately, when it comes to the common metric–imperial conversions most involve little more than doubling and halving. We used the table overleaf in *Maths for Mums & Dads*, but it's so handy that we've included it again.

* We won't allow songs like 'How Can I Metre?' by the Everly Brothers.

Conversion	Very roughly	More accurate
Inches to centimetres	Double	Multiply by 2.5
Yards to metres	The same!	'A metre measures three foot three, it's longer than a yard you see.' Taking off 10 per cent to turn yards to metres is close enough for most everyday situations.
Miles to kilometres	Double	Multiply by 8 and divide by 5
Fahrenheit to Celsius	Take away 30 then divide by 2	Take away 32 then multiply by 5 and divide by 9
Gallons to litres	Double twice	Multiply by 4.5
Pounds to kilograms	Halve	'Two and a quarter pounds of jam weigh about a kilogram'

A neat conversion from miles to kilometres

Many children learn about a sequence of numbers that runs like this:

1 1 2 3 5 8 13 21 34 55 . . .

Each number in the sequence is found by adding the two previous numbers. It is known as the Fibonacci sequence, and is famous for the fact that these numbers often crop up in nature. For example, the number of petals on a flower will frequently be a Fibonacci number.

But the sequence has another quirky feature. Any neighbouring numbers in the Fibonacci sequence represent miles and their equivalent in kilometres. So 5 miles is 8 kilometres, 13 miles 21 kilometres, 21 miles is 34 kilometres and so on. And this is uncannily accurate. Why? For reasons that are quite subtle, and beyond the maths of this book, the ratio of neighbouring numbers in the Fibonacci sequence is very close to something known as the *golden ratio*, a number slightly larger than 1.61. And by pure coincidence 1.61 is also the ratio of kilometres to miles.

Anyway, next time you are driving in France, keep the Fibonacci sequence handy and when a road sign indicates 55 kilometres and somebody asks what that is in real money, you can confidently announce that it is *almost precisely* 34 miles. (If the distance is not a Fibonacci number, you may have to resort to: 'Hang on for another couple of kilometres and I'll be able to give you a precise answer.')

Rounding, and 'ish' Numbers

Rounding crops up in many areas of maths, but never is it more relevant than in measuring, since builders, shopkeepers, statisticians and just about any other professional that uses numbers deals with rounding all the time. There's no rule about how accurately a number needs to be rounded to, as it depends so much on the situation. Weighing out flour for a cake, it's usually fine to round to the nearest 100 grams. On the other hand if you are measuring a child's temperature, you need to be accurate to one decimal place. A temperature rounded to '38° Celsius' could mean anything from a mild temperature (37.5 °C) to a fever (38.4 °C).

Numbers can be rounded to any place value, for example to the nearest hundred or to the third decimal place. Remember, digits smaller than 5 are rounded down, digits that are 5 upwards are rounded up. For example 7.3 rounded to the nearest whole number is 7, while 84.5 is rounded to 85.

If a series of rounded numbers are added together, there is a chance that the total could contain a significant error. For example, if you round before adding 2.5 + 3.5 + 4.5 + 5.5 then the total of 18 is quite a way from the correct answer of 16. There's a neat way of reducing this error, using a method known as 'Bankers' rounding' in which you always round to the nearest even number. 21.5 would be rounded to 22, whereas 20.5 is rounded to 20. The sum of 2.5 + 3.5 + 4.5 + 5.5 rounds to 2, 4, 4, and 6 and, hey presto, the total is 16! (It isn't always quite this neat.)

TEST YOURSELF

A carpet fitter is checking the area of a floor in a rectangular room, he takes accurate measurements and then rounds them to the nearest 10cm. He reports back that the room is 4.3 metres × 3.1 metres, or 13.3 square metres. How much larger than 13.3 m^2 might the actual area of the floor be?

Compound Measures

The name 'compound measures' is a bit of jargon that has entered school maths language. In simple terms it means those measures that are a combination of at least two units, such as distance and time (which make speed). You can also tell something is a compound measure if its name includes the word 'per'. For example, density is 'kilograms per cubic metre', while speed is 'metres per second' (or 'miles per hour'). Not all measures are so obviously compound. What about price? It sounds like a straightforward measure, but when we say price, we actually mean 'cost per unit'. And temperature? This is actually the amount of heat per unit of volume.

The reason why we're making a point about this is that

teenagers tend to think that compound measures can be added and divided in just the same way as simple measures, and make many mistakes as a result.

If you have a 40-litre bucket of water and add it to a 60-litre bucket, what is the combined number of litres? 100 litres, of course. But what if the temperature of water in the first bucket is 40 °C and the second is 60 °C? Add the two buckets together and the combined temperature of the water is . . . ? Common sense should tell you that it *isn't* 100 °C, since that would mean you have miraculously created boiling water without adding energy, though when we asked a sample of teenagers this question a surprising number did give that answer. The answer for the combined buckets is that the temperature will be somewhere between 40 °C and 60 °C, the exact temperature depending on how much water is at each temperature. (In this case it actually works out as 52 °C.)

In the same way, careful thought is needed before you combine two speeds or two pressures or two densities.

TEST YOURSELF

Imagine there is a road that goes up a hill for 10km, and down the other side for another 10km. Two friends, Rita and Peter, decide to take a trip cycling up the hill and down the other side. Rita does the whole journey, up and down the hill, at 10km per hour, so it takes her an hour to get to the top and another hour down the other side. The whole journey takes her two hours. Peter isn't as fit and cycles up the hill at only 5km per hour. How fast does Peter need to cycle down the hill in order to catch up with Rita before she reaches the bottom of the hill? Your choices are:

a) He needs to cycle down at 15 km/hour so that his average speed is the same as Rita's $(5 + 15) \div 2 = 10$.
b) Since he cycles up at half Rita's speed, he has to make up for it by cycling down at twice her speed, in other words he has to cycle down at 20km/hour.
c) It isn't possible for Peter to catch Rita.

Temperature

Temperatures in the UK are almost always now quoted in °C, where C these days stands for Celsius (he was the man from Sweden who invented the scale in 1742) and also for its earlier name, Centigrade, which some of us still use out of habit. Centigrade hints at the fact that the temperature is measured on a scale of 0 (freezing point of water) to 100 (boiling point). Most Americans and some British prefer to use Fahrenheit, which has more awkward numbers for water (32° for freezing and 212° for boiling) but does have the advantage that the number scale gives us nice ranges to describe the human condition. The tabloid headline back in 2003 that said: 'Britain sizzles in 100° heat for the first time' would have had rather less impact if it had said 'Britain sizzles at 37.8'. Parents generally know to start getting concerned if a child is running a temperature above 100, but it's harder to remember where that feverish temperature begins in its Celsius equivalent.

The formula for converting Celsius into Fahrenheit is a good example of a linear equation: Fahrenheit $= \frac{9}{5}$ Celsius $+ 32$ (or $F = 1.8\,C + 32$).

There's a rule of thumb that makes this conversion much easier, by removing that awkward ratio of 1.8. To convert from

Celsius to Fahrenheit, simply double and add 30. For most of the temperatures used in weather forecasts this is accurate to within a degree or two, and for 10 °C it gives precisely the right answer (double 10 and add 30 gives 50°, and 10 °C is indeed 50 °F).

The reason why this approximation is so accurate becomes clearer if you plot the graph of the exact formula and the approximate one. The two lines run extremely close to each other in the range of outdoor temperatures we usually deal with:

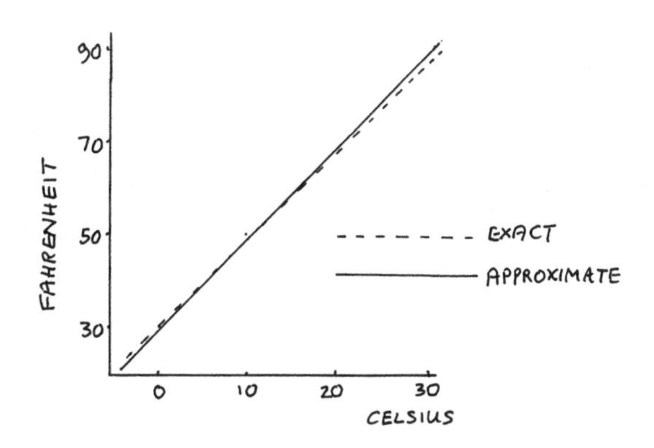

TEST YOURSELF

1) What is the reverse formula, for converting from Fahrenheit into Celsius?
2) Harder question: at what temperature are Fahrenheit and Celsius the same?

Perimeters and Area

The formula for area that teenagers first become familiar with is for calculating the area of a rectangle: length times breadth. If L stands for the length of one side of the rectangle and B the other, then area can be expressed as $L \times B$. And the perimeter of the rectangle is $L + B + L + B = 2(L + B)$.

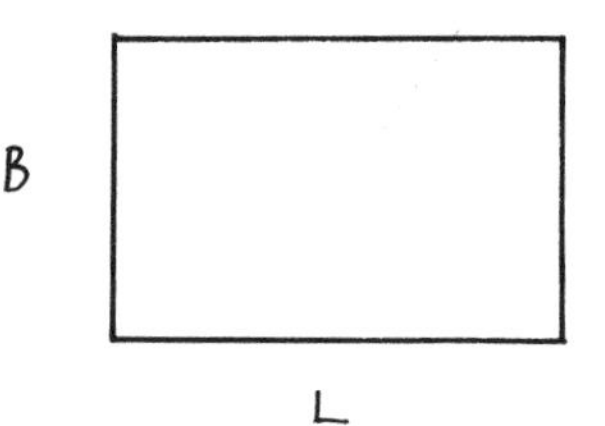

If you increase L or B then the perimeter gets bigger, and so does the area. However, that can lead to the faulty conclusion that it is always the case that the longer the perimeter the greater the area and vice versa.

A loop of string is all you need to explore how the relationship between area and perimeter is not that straightforward. The loop of string can be laid out to make a rough square but it can also be arranged to make a long thin rectangle – and the area of the rectangle is clearly much less than the area of the square. The perimeter (the length of the string) is fixed but the area changes (in fact, the string surrounds the greatest area when it is a circle).

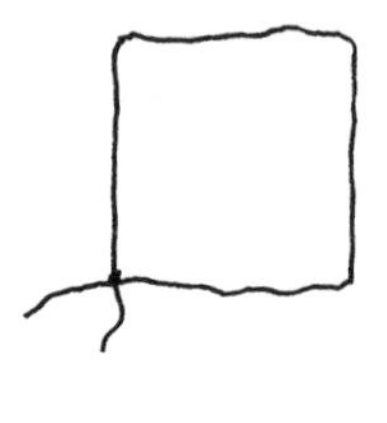

Most of the problems your teenager will meet around area and perimeter will involve shapes that can be divided up into a number of smaller rectangles. Usually not all the lengths of sides are given, but the detective work to figure them out is not too baffling.

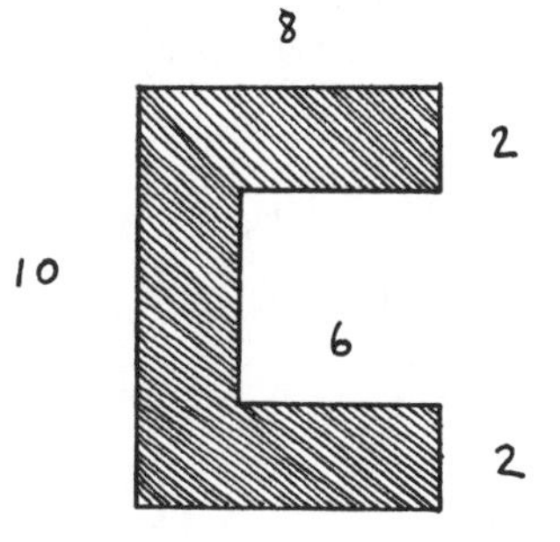

Calculating the perimeter of this polygon involves seeing that the bottom edge must also be 8 units and the short interior vertical edge $10 - 2 - 2 = 6$ units. The calculation is a simple adding up: going clockwise around from the top: $8 + 2 + 6 + 6 + 6 + 2 + 8 + 10 = 48$.

We could calculate the area by carving the shape up into rectangular pieces and calculating the area of each.

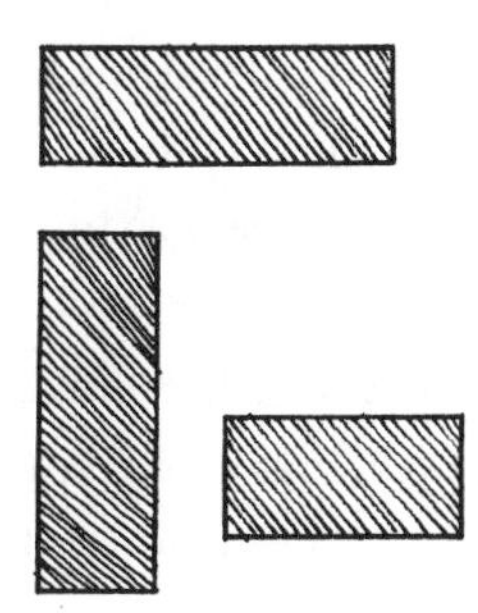

But it's quicker to see the shape as a rectangle with a piece cut out.

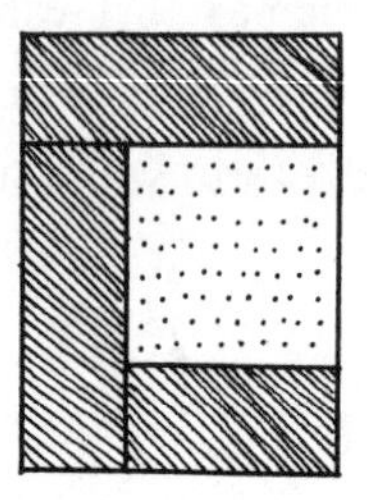

The large rectangle is $10 \times 8 = 80$ square units, the gap $6 \times 6 = 36$ square units, so the area of the polygon is $80 - 36 = 44$ square units.

Circles and Pi

Perimeters of shapes are relatively easy to work out, at least when they involve straight lines. It requires little more than adding up. But things move up a level when it comes to circles. Most people remember that circles involve the number 'pi', though beyond that, memory tends to be hazy. As a reminder, pi (the Greek letter π) is the ratio of the perimeter of a circle (usually known as the *circumference*) and its diameter. Teenagers will often encounter it for the first time through an experiment. Cut a piece of string so that it is the length of the diameter of a circle – the top of a baked-beans tin, for example. Then see how many lengths of that string it takes to surround the whole circle. The answer is just over three. A more accurate answer is 3.1415926, though even that is only an approximation, since the number itself cannot be represented as a finite number of digits.

The formula for the circumference of a circle is written in

two ways: πD where D is the diameter of the circle, or more commonly (since the diameter is double the radius) $2\pi R$.

The equation for the area of a circle is similar: $A = \pi R^2$ (pi multiplied by the square of the radius). The fact that πR crops up in both equations means that teenagers invariably get them mixed up when they try to recall them.

One common-sense way to remember which formula is which is that equations about area are the ones that will have 'squares' in them. In fact, there is a rather elegant way of showing where the formula for a circle's area comes from.

Imagine a circle as a pizza, radius R, sliced into eight wedges. You can rearrange the wedges to make a shape that's quite rectangular:

Perimeter of circle is $2\pi R$

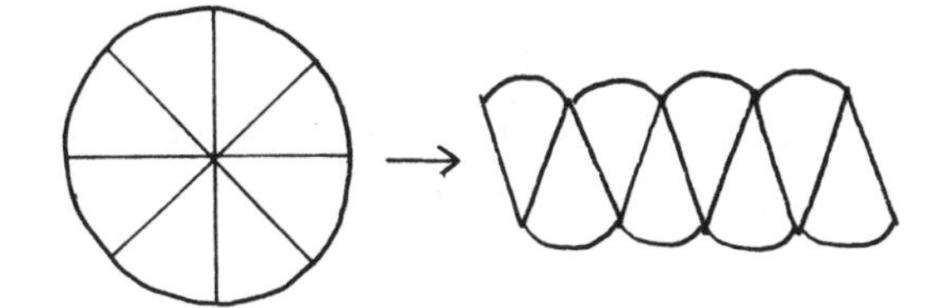

What if you now cut the pizza into incredibly thin wedges, hundreds of them, and reassemble those wedges?

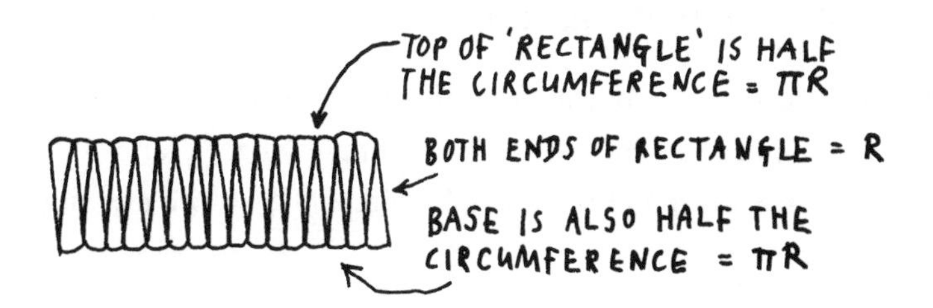

Now the new shape really does begin to resemble a rectangle. The two ends of the rectangle are both simply the radius R of the pizza, while the top and the bottom sides together make up of the entire crust of the pizza – in other words, the base of the rectangle is half the circumference, so it is half of $2\pi R$ which is πR.

So the area of the rectangle (and the circular pizza from which it was made) is its length times its height, $\pi R \times R$, or πR^2!

TEST YOURSELF

a) What is the difference between the circumference of a circle with a diameter of 2 metres and another whose diameter is 4 metres?

b) The curves at the end of an Olympics athletics track are semicircles, and each running lane is about one metre wide. In the final lap of the 800 metres at the 2004 Olympics in Athens, the front runner ran the whole lap in the inside lane, while Britain's Kelly Holmes ran the lap in the second lane in order to overtake the other athletes and win the gold medal. Roughly how much further did Kelly Holmes run on that final lap?

Mixed up with mnemonics

Many teenagers are encouraged to memorise mathematical rules using mnemonics, including, for example, the equations for the circumference and the area of a circle. These two are quite common:

Cherry Pies Delicious (Circumference = Pi × Diameter)
Apple Pies Are 2 (Area = Pi × Radius squared, or πR^2)

These might be OK if remembered exactly. Unfortunately mnemonics in maths are often only half-remembered, and many of the ones we've seen do little to encourage understanding.

We met one quite able teenager who recalled the mnemonics above as: 'Cherry Pies Are Delicious', and then, hearing herself say 'Pies Are', which sounds like 'Pi R', translated this to mean Circumference = Pi × Radius × Diameter. Meanwhile, 'Apple Pies Are 2' sounds rather like 'Pi R 2', which itself is like '2 Pi R' – the formula for the circumference. Having got herself tied in knots, she had to look up the formula so that she could get the right mnemonic, which of course made the whole idea of the mnemonic pointless.

Volume

In primary school, children investigate volume using measuring jugs. In secondary schools, they are expected to calculate it – not least because you can't use a measuring jug to find the volume of, say, a house. Volume is usually measured either in cubic metres or in litres (the latter is the normal measure for liquids, of course), but teenagers are expected to convert between these different measures.

Here's a reminder of the different standard volumes (drawn to different scales!):

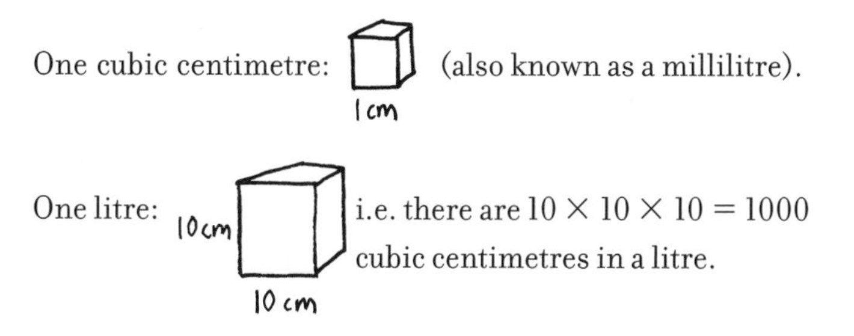

One cubic centimetre: (also known as a millilitre).

One litre: i.e. there are 10 × 10 × 10 = 1000 cubic centimetres in a litre.

One cubic metre:

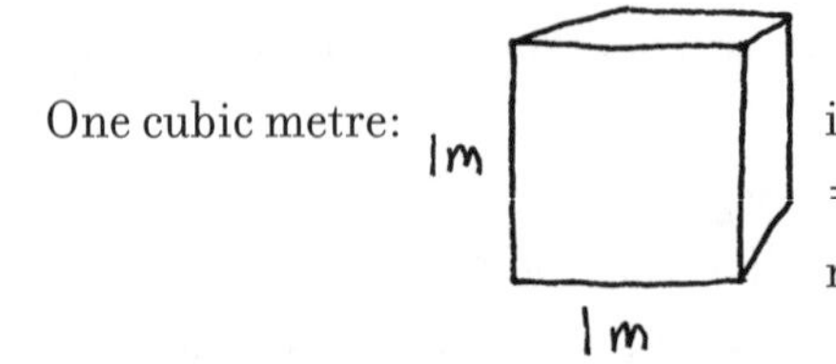

i.e. there are $10 \times 10 \times 10 = 1000$ litres in a cubic metre

Inevitably teenagers are asked to convert between these different units, and inevitably they make mistakes. In the case of litres to cubic metres, for example, it's common to divide by 10 instead of 1000 (because each side of the cubic metre in the diagram is ten times longer than in the litre cube).

Indeed, this mistake is made in comparing most volumes. If you have a box of volume 15 litres and double the length of each side, what is the volume of the larger box? The most common answer is 30, whereas the correct answer is that the box is $2 \times 2 \times 2 = 8$ times larger, so the larger box has a volume of 120 litres.

With bigger conversions, the number of zeroes causes the problem. Most measurements that your teenager will encounter will have one of these prefixes:

Kilo	1000
Centi	$\frac{1}{100}$th
Milli	$\frac{1}{1000}$th

Surface Area and Volume

Like areas and perimeters, volume and surface area are often studied together. If you think of volume as the contents of a parcel, then surface area is the amount of wrapping paper needed to cover the outside. As the volume of the box increases, you might expect its surface area to increase at the same rate: if you double the volume, you might expect to double the surface area. But this is not the case.

The easiest way to discover this is to take a simple cube with sides of 10cm. Its volume is 1000cm^3. It has six faces, each of 100cm^2 so its surface area is 600cm^2 . If you double the length of the sides, its volume is now $20 \times 20 \times 20 = 8000\text{cm}^3$, while its area is now $20 \times 20 \times 6 = 2400\text{cm}^2$. The volume has increased eightfold while the area has only increased fourfold – the ratio of the volume to surface area has got larger as the object grows. In fact, it is true for any solid object that the larger the dimensions, the larger the volume to surface area becomes.

In the case of 'prisms' (anything from a cylinder to a triangle prism like a Toblerone® box), surface area and volume are closely connected. The volume of a prism can always be calculated by multiplying the area of the end shape by the length of the prism, which we'll call L.

So the surface area of a triangular prism is the area of the end triangle ($\frac{1}{2}$ base $\times$ height)$\times$ L. And the volume of a cylinder is $\pi R^2 L$.

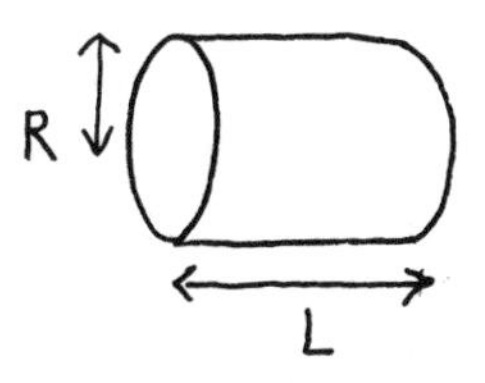

Surface area and evolution

The relationship between volume and surface area has been a vital factor in the evolution of species.

For any 3D shape, such as a cube or a sphere, the ratio of the surface area to the volume gets smaller as the size increases. The graph shows this for a cube:

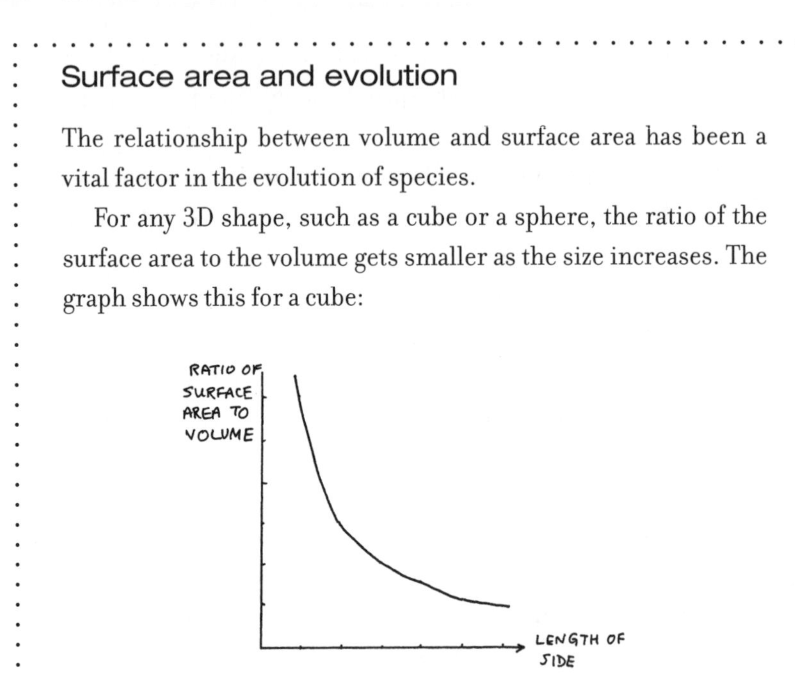

Small creatures have extremely thin legs because the small area of their feet is enough to support the pressure from their weight. However, as the volume of creatures increases so does their weight. To support that weight, the creature's feet have to be disproportionately larger, hence the huge feet of elephants. One reason why classic science fiction B-list films like *Them! (Attack of the Giant Ants)* will never become a reality is that the ants' spindly legs would be unable to support their now giant bodies, and they would collapse into a writhing heap.

TEST YOURSELF

a) How many centilitres are there in a litre?
b) How many cubic centimetres are there in a cubic metre?
c) What's the most cubes with 20cm sides that could be fitted into an 800-litre vault?
d) If a cylinder has circular ends of radius z and height a, what is its volume, and what kind of cylinder is it?

Measurements and Algebra

One reason why volume and area get particular attention in school maths is that they are a way to link geometry and algebra. We've seen earlier that the perimeter P of a rectangle is:
$P = 2L + 2B$.

While the equation for its area is $A = L \times B$.

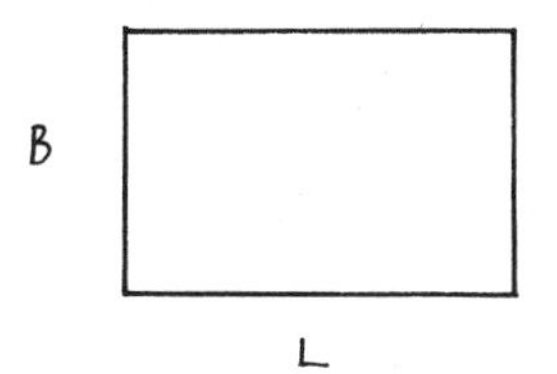

Problems involving area become more interesting – and more challenging – when you are trying to find a 'best answer'. Imagine you have a square sheet of card, 10cm × 10cm. You want to turn this into a tray by folding up the sides, and you want the tray to have as large a volume as possible. (If you like a physical example, imagine you want the tray that holds as many paperclips as possible.)

To make a tray, cut out squares from the corner and call the side of these squares S:

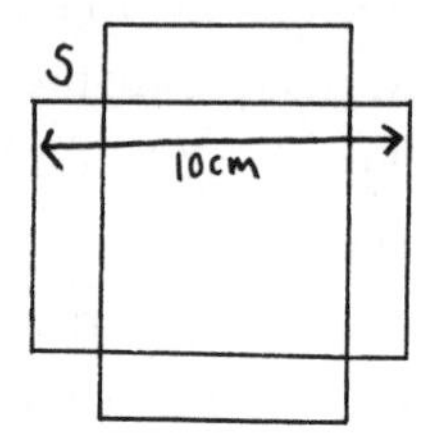

How big should S be if the box is to hold as many paperclips as possible? If S is very small, you have a box with a large square base but almost no height, so its volume will be tiny. On the other hand, as S approaches 5cm, you end up with a skinny tube that is nearly 5cm tall but has hardly any base, so again its volume is tiny. There must be somewhere in between those extremes where the volume is the maximum, but it's certainly not obvious where that will be.

One way to find out is to produce a formula for calculating the volume of the box.

- The square base has sides that are (10cm − $2S$) long, so the area of the base is $(10 - 2S)^2$.
- The height of the box is S.
- The volume is the area of the base multiplied by the height: $(10 - 2S)^2 \times S$*

To get an impression of how the volume varies, you can produce a table. Better still, set up a spreadsheet, and then draw a graph of how the volume changes as L increases:

* Multiplied out this becomes the slightly scary: $4S^3 - 40S^2 + 100S$.

S (cm)	Area of base (cm^2) $= (10 - 2S)^2$	Volume of the box (cm^3) $= S \times A$
0	100	0
1	64	64
2	36	72
3	16	48
4	4	16
5	0	0

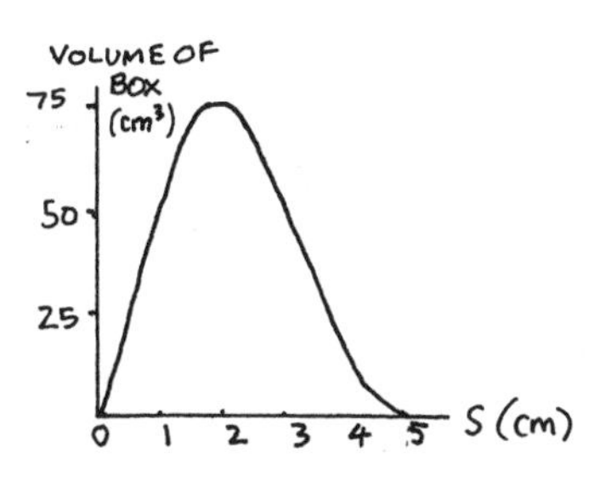

Notice how the volume of the box starts at zero when the side of the square is zero (there's nothing to fold up!), then builds up to a maximum value somewhere between $S = 1$ and $S = 2$ cm, before falling away to zero again when the squares cut away the entire sheet of card. A spreadsheet can easily be adapted to make the increments of S smaller, and if you do this it turns out that the peak of the graph is at roughly $S = 1.7$. In fact, the precise value for which the box has the maximum volume is when the length S is exactly $\frac{1}{6}$ the length of the original large square of cardboard (in this case $\frac{5}{3}$ cm, or 1.6666cm). If you wanted to prove this result, you'd need to use *calculus*, maths that is beyond regular GCSE. But you don't need calculus to be able to explore problems like this and to get a feel of how volume and algebra are related.

MEASUREMENT
If you do only three things . . .

- Try the 'imperial and metric' quiz on your teenager: they'll probably enjoy guessing George Clooney's height, and you can use the answers as a starting point for metric conversions.
- Encourage your teenager to identify measurement 'benchmarks' to judge answers against. For example, a door is about 2 metres tall, a medicine spoon holds 15ml. And casually switch between metric and imperial (the 2-metre door is a bit more than 6 feet, the height of somebody tall).
- Discourage your teenager from relying on mnemonics for the area and circumference of a circle, they are too easily confused. The best way to learn them is by practising. As a reminder of where the circumference formula comes from, do the experiment of wrapping string around the circumference of a tin and seeing how many diameters long it is.

CALCULATORS

'In my day, we didn't have calculators . . . Actually we did, but they weren't nearly as fancy as the ones they have now.'*

There are some who regard the calculator as the root cause of everything that has gone wrong in maths education in the last few decades. Others, though, see the calculator as the tool that has liberated children from much of the drudgery of maths, enabling them to do serious problem-solving, and even enjoy the beauty of the subject, without getting bogged down in the arithmetic. Inevitably the truth lies somewhere in between.

Whatever your view of calculators, they are a major, unavoidable feature of secondary-school maths, which is why we've given them a chapter all of their own. Every child will be expected to have a calculator, and examiners will expect them to know how to use one for at least some of the exams.

* Reminiscing parent.

However, let's not get carried away with the importance of calculators in maths. A student studying maths at university might never use a calculator, at least if they are studying 'pure' maths. Maths (as opposed to arithmetic) is more about reasoning and problem-solving than it is about calculating. That's one reason why GCSEs always have a non-calculator paper, and even in the paper that *does* permit calculators, the device isn't nearly as helpful as you might expect – particularly if, as is the case with many teenagers, you don't know how to use it properly.

Some schools told us that their pupils score worse in the calculator papers than in the non-calculator papers. Why? Partly because pupils accidentally press the wrong buttons, putting a decimal in the wrong place or keying digits in the wrong order and have no idea that they've made an error. Another reason is that often the calculation part of a question is relatively straightforward: it's knowing how to use the functions and which buttons to press that is the hardest part. And the sad fact is that many pupils aren't skilled at using calculators.

Tips for Calculators

1) Buy the one that your school recommends. It means this will be the calculator your teenager's teacher will be familiar with, and will save those awkward moments when the teacher asks the class to use a button that has a different label from your child's.
2) Use the same calculator in exams as you use in school and at home – and make sure it's working before the exam, and has a fresh battery. There have been many panics caused by pupils forgetting calculators for exams and having to borrow an unfamiliar one on the day.

3) Never use the percentage button. (See *The Scientific Calculator* below.)

What Sort of Calculator?

There are three types of calculator used in schools:

1) **The basic calculator.** This is the type you can get from a Poundshop. It's cheap, cheerful and with its arithmetic functions and a square root button, it's practically all you need for everyday life. It also has the sort of old-fashioned display that allows you to type in 71077345, turn it upside down and get the word SHELLOIL. Ah, we used to make our own entertainment . . .

   However, a basic calculator will be detrimental to your child's secondary-school maths, not only because it lacks many important functions your child will encounter (brackets, trigonometry, etc.) but also because its logic can be back to front. To find the square root of 64, for example, instead of entering [√] [6] [4] you have to enter 64 first and then enter the square root. And these calculators are hopeless when it comes to the conventions of which order to do a calculation such as: $3 + 8 \times 4$. A basic calculator will tell you the answer is 44, whereas any maths teacher can tell you the correct answer is 35, because the convention is to do the multiplication before the addition. (See BIDMAS on page 257).

2) **The Scientific Calculator.** Secondary schools and exam boards expect every child to have one of these. They have all the functions that you'll ever need for maths GCSE and

A level, plus several buttons you'll never use at all. In fact, because these calculators are enough to see most university students through their studies, scientific calculators are arguably too complicated for some GCSE pupils, who may get confused by the vast array of functions and buttons. The most important feature to look for in a scientific calculator is a display with two lines, one to show the workings and the other to show the answer.

One button that most scientific calculators don't have is a % button (and even if they do have one, it's usually hidden away somewhere). This is because % buttons are a curse. Unless you are very familiar with exactly how the one on your particular calculator works, the chances are you'll struggle to do a calculation using it. Typically the only people who can successfully use the % button on a calculator are those who already understand how percentages work, in which case they are less likely to make a mistake if they just work out percentages 'long hand' using the other calculator buttons, in particular by dividing and then multiplying by 100.

3) **The Graphics Calculator.** With larger displays and additional function buttons, these calculators go a step further and allow you to plot graphs and investigate properties of shapes, a great visual aid for those who can get to grips with operating them. Many schools now teach sixth-formers to use graphics calculators. But it's unlikely your teenager will use one of these, and if they do, the maths they are doing is probably beyond the scope of this book.

BIDMAS and BODMAS

Some call it BIDMAS, just as many call it BODMAS, and you'll also hear it referred to as BEDMAS, BOMDAS and various other things, some of them unrepeatable. Whatever your teenager's school calls it, this is the rule that reminds you the order a calculation in which be done should. Or put another way, the order in which a calculation should be done. You see, the order really does matter.

The letters stand for:

Brackets
Index (or Order, or Exponent)
Division
Multiplication
Addition
Subtraction

In longhand this rule means that you should calculate the things in brackets first; then work out any numbers that have been raised to a power (otherwise called an index or exponent); then do the divisions and multiplications; and finally do any remaining additions and subtractions, starting from the left. For example, calculate: $7 + 3 \times (9 - 5)^2$.

Using BIDMAS:

- Work out what's in the brackets first: $9 - 5 = 4$
- Now do the exponent: $4^2 = 16$
- Now multiply: $3 \times 16 = 48$
- Finally add: $7 + 48 = 55$

This is a useful guideline, and it's unlikely that in GCSE maths your teenager will encounter any examples where BIDMAS might lead to problems. Calculators use these rules, too, so if you enter the calculation above on a scientific calculator, it should also come up with the answer 55. (If it doesn't, either you've made a mistake or you should make sure your teenager uses a different one.)

Be warned, however, that there *are* some grey areas with BIDMAS that serve as a reminder that this is a handy aid but is not the definitive rule. For example, every so often, the following calculation is discussed on Facebook or Twitter: $6 \div 2(2 + 1)$.

Notice that there is no multiplication sign used here: the convention is that a number written before brackets means 'multiply this number by the number in brackets', so that $2(2 + 1) = 2 \times 3 = 6$.

So what does BIDMAS say the answer is to $6 \div 2(2 + 1)$?

First the brackets: $2 + 1 = 3$.

Now what? According to BIDMAS you should now 'work from the left', so $6 \div 2 = 3$, then multiply that by $(2 + 1)$ to give $3 \times 3 = 9$.

However, ask most mathematicians, and they will tell you that there's an understanding that if you leave out the multiplication sign, you should do that multiplication before doing any division. They will argue that the calculation above is ambiguous, but that if it means anything at all, then it should be interpreted as $6 \div (2 \times (2 + 1)) = 6 \div 6 = 1$. And to prove that this isn't just an argument between people, the authors have two scientific calculators, different models manufactured by the same well-known company, and one calculator gives the answer as 9 while the other says it is 1. If scientific calculators can't agree, you know you are in trouble!

So just remember, BIDMAS is a handy rule, but it's not the *definitive* rule.

How Scientific Calculators Have Changed

In many of their features, scientific calculators haven't changed in the last thirty years. Sines, cosines and square roots produce the same result today as they've always done, just like a Hovis loaf. However, beware, there are some important changes and refinements that can catch out an unprepared parent. They've even invented a couple of new buttons.

	1980s scientific calculator	2010s scientific calculator
The display	When doing a calculation, there was only ever one number on the display. This was the last number you entered or calculated. For example, if you calculated $4^2 + 8$, what you would see on the display would be 4, then 16, then 8, and when you pressed [=], the final answer 24.	All the entries of the calculation are kept on the display, so $4^2 + 8$ appears in full, exactly like that. When you press [=] the answer is displayed on the line below so you can now see the entire calculation and its answer. This is *much* better, as it makes it easy to look back at what you've done, and check you haven't made an error when entering the calculation.

	1980s scientific calculator	2010s scientific calculator
Deleting	There were two buttons, [C] to clear the number on the screen, and [AC] to clear everything and start a calculation again.	[C] has now been replaced by [DEL], a delete key similar to one on a computer keyboard that deletes the last thing you entered on the display. So [3] [4] [5] [DEL] will delete the [5] leaving [3] [4]. Most calculators also have left/right and up/down cursor keys, that allow you to move around the display and edit an error. Much more user-friendly!
Symbols	The calculator was only able to display a limited number of symbols (such as '+' or the degree symbol 'º').	Most are able to show fractions, square root symbols and indices, which means that what you see on the display looks similar to what appears in the textbook.
Memory buttons	Had prominent Memory buttons, such as M-In and M-Recall. Often you did long calculations in steps, saving the answer in the memory and then recalling that answer to use in the next step of the calculation.	The Memory button is still there, but it is less prominent and pupils hardly ever use it. Instead they use the ANS button (which effectively means 'Use the last answer in this next calculation'). Doing a two-step calculation, the first step might produce the number 21.3 (say), then if the next step is to double that number, you press [2] [X] [ANS].

	1980s scientific calculator	2010s scientific calculator
Multi-plication	If you wanted to multiply, you ALWAYS had to use the multiplication button. To work out 8sin(30) you had to enter 8 [×] sin(30) or you'd get an ERROR message.	In expressions where a multiplication symbol normally isn't used, for example 8sin(30) or 3(7 +4), the calculator will allow you to leave out the multiplication symbol. So you can enter either 8 [×] sin(30) or 8sin(30) and both should give you the correct answer, which is 4.
Function buttons	If you wanted to calculate the square root, sine, cosine, etc., of a number, you entered the number first and then pressed the function button. So to find the square root of 64 you pressed 64 [√].	These days you press the function button first, then the number. So [√] [64] [=] gives the answer [8], and [Sin] [30] [=] gives [0.5]. This feels much more natural, as it uses the terms in the same order as they'd be used in a textbook.
Reciprocal button	To make a fraction such as 1/5 you entered [5] and then pressed [1/×] to get the answer 0.2. The [1/×] button also enabled you to invert any fraction, so [2] [/] [3] [=] gave 0.666 . . . and pressing [1/×] gave you 1.5, which is 3/2.	The button hasn't changed, but its label has: it is now the x^{-1} button. For those going on to higher maths this is good notation, but for teenagers lacking confidence in maths, it has turned the function into more of a black box. What has 'x-1' (as they see it) got to do with dividing a number into 1?

	1980s scientific calculator	2010s scientific calculator
The [S⇔D] button (sometimes the [F⇔D] button)	Didn't exist!	This button takes the drudgery out of simplifying fractions and converting them to decimal numbers. What's slightly confusing is that S⇔D stands for surds to decimals, because this function converts *surds* (see the Glossary) as well as fractions. As an example enter [7] [÷] [1] [8] and most calculators will give you the calculation *as a fraction: 7/18.* Press [S⇔D] and you get 0.3888… And if you enter [6] [/] [8], the calculator simplifies this to ¾ , and the [S⇔D] button will then tell you this is 0.75. Very handy . . . though it means many pupils have lost the knack of simplifying and converting fractions in their heads.
Format of the digits	When you entered the number 71077345, and turned it upside down you saw SHELLOIL.	The digits today are properly formed and nicely rounded. It means that instead of 'SHELLOIL' you now see this: **71077345** (printed upside down) And they call this progress?

CALCULATORS
If you do only three things . . .

- Remind your teenager to use Zequals (page 41) to calculate an approximate answer and use this to check the reasonableness of the calculator answer.
- Encourage your teenager to check calculator answers by repeating the calculation and, if possible, changing the order of calculation – if the answer is different, find another way.
- Buy the calculator that the school recommends.

PROBABILITY AND STATISTICS

It has been estimated that approximately 72 per cent of statistics have been made up on the spur of the moment.

Probability is the maths behind a lot of what people do for fun. The tactics in many of the most popular games, from poker to Monopoly, depend on it. A whole industry in the form of TV gameshows has also been built on chance: the people who devised Bruce Forsyth's *Play Your Cards Right* and more recently *Million Pound Drop* cleverly created rules to ensure that with some luck, a few but not too many contestants will win the jackpot. And winning the National Lottery is of course all about probability, albeit an extremely tiny one.

More seriously, probability is also the maths behind everyday decision-making. Should I bother to get health insurance? Should I put my money in a building society or in shares?

Given all this, it is quite surprising that a mathematical topic so relevant to everyday life was barely taught in schools until the 1980s. What it means is that many parents and almost all grandparents had no formal education in probability at all.

Today, our world seems to be driven by statistics. Just about anything that can be measured is measured, and hardly a news story goes by that isn't based on a statistic of some form. So it isn't surprising that statistics feature so prominently in school – though there are many that argue that, as a topic, it belongs just as much in the sciences, geography and economics as it does in maths lessons. (And indeed some teenagers do a separate GCSE in statistics.)

PROBABILITY

Many teenagers instinctively make decisions that show they 'get' probability without any formal understanding of the maths. For example, in a game of *Play Your Cards Right*, when you have to guess whether the next card turned over will be higher or lower, almost every thirteen-year-old when shown the five of Hearts will guess that the next card turned over will be of higher value (as indeed it will be, about $\frac{3}{4}$ of the time).

The problem begins when working on a hunch is no longer enough, and you need to actually calculate the chances. When probability becomes mathematical, most teenagers have difficulties. There are plenty of pitfalls for the unwary, and there are times, particularly when we encounter coincidences or assess relative risks, when human intuition gets probability completely wrong.

The Language of Probability

Children first encounter probability in primary school, where it is presented in the form of words, ranging from 'Absolutely certain' through 'Likely' then 'Unlikely' and finally 'Never' (or zero). What is the chance that the sun will rise tomorrow? Certain. That a hare would beat a tortoise in a race? Very likely. That Preston North

End will win the FA Cup next year? Unlikely. That the Prime Minister will appear on the news tomorrow dressed as a clown? Almost zero.

The next step, in secondary school, is to attach numbers to these rather vague words. 'Certain' means if you could test something, say, 1000 times, the stated outcome would happen 1000 times, or 100%. Zero chance means it would happen zero times, which is 0%, of course. So probabilities are expressed in the range 0% (will never happen) to 100% (it will always happen). Anybody heard to claim 'I'm 1000% confident' is using a bit of hyperbole, just like the footballers who claim to have given it '110 per cent' or indeed the Beatles whose love extended to 'eight days a week'.

Things get more interesting, and challenging, with events that are somewhere between certain and uncertain, such as tossing a coin. Everyone knows that when tossing a coin, there are two possible outcomes and that the chance of getting a head if you toss a fair coin is 'fifty-fifty'. Its probability is 50%.

However, many teenagers assume that 'fifty-fifty' will apply to any situation where there are two possible outcomes. When asked the question: 'What is the chance I will break my leg tomorrow?', the majority of thirteen-year-olds answer 'one half', on the basis that there are two possible outcomes, either I break my leg or I don't. Once they apply some common sense, they realise, of course, that this is nonsense, but sometimes common sense is lost when trying to represent real situations as numbers.

Five other ways to express a chance of '50%'

A fair coin, when tossed, will land heads up half the time and tails the other half, so the chance of a head is 50%. However, probabilities aren't always expressed as percentages. There are several ways of saying the same thing. The chance of getting a head on a coin is also:

1 in 2
½
Fifty-fifty
0.5
One-to-one (or 'evens' if you are a bookmaker)

Which should you use? It depends on the circumstances. Percentages are 'friendlier' (weather forecasters tend to use them when indicating the chance of rain) but if you are calculating the chances for a combination of events – for example, throwing a head followed by a tail on a coin, or the chance of it raining tomorrow AND of the train being late – then it's normal for mathematicians to use fractions that can then be multiplied together.

In school maths, fractions are the most common way to write a probability. For example, the chance of rolling a 5 (say) on a dice is 1/6.

Working out the Probability of an Event

How do you *know* that the chance of rolling a 5 on a normal dice is '1 in 6'? How do you know any probability, for example, the chance that Manchester United will win their next match, or that it will rain tomorrow?

The key to working out probabilities is to list all of the possible outcomes. In the case of tossing a coin, there are only two outcomes, a head or a tail, but with something like the weather there are several possible outcomes – rain, snow, sunshine, overcast, and so on. Once you know the outcomes, you then need to know the proportion of times that each of these outcomes is expected to happen. This is the probability.

There are five different ways in which probabilities are calculated (or estimated).

1) **Look at the relative proportions.** A dice has six identical faces, and its symmetry means there is no reason why any of the faces will come up more often than any other. So the chance of a particular face cropping up next will be $\frac{1}{6}$. However, if you have (say) a spinner and it is divided into three portions of different sizes, then the portions are not equally likely to come up.

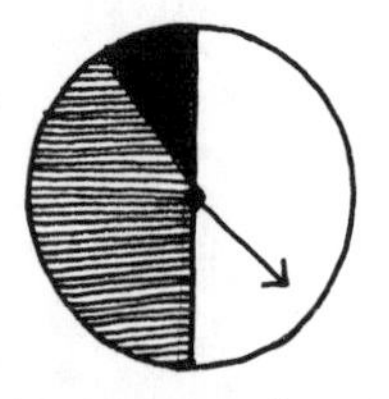

For example, in this spinner, the circle is divided into three regions. The white region (from noon to six o'clock) is half of the circle, the shaded region (from six to eleven o'clock)

is 5/12 of the circle, and the solid black region (from eleven until noon) is 1/12 of the circle. So if you spin the arrow around, then it's reasonable to assume that the chance of it ending up pointing to the white, shaded or black regions are ½, 5/12 and 1/12 respectively.

2) **Find out by experimenting.** If you toss a toy pig in the air, what is the chance it will land on its feet. Who knows? Its shape is too complicated to be able to use symmetry or proportion to work it out. The best way to find out is to experiment by tossing the pig a few times. If you toss it ten times and it lands on its feet once, then 1/10 is an estimate of the probability it will happen next time. However, be careful with a small sample like this, as there will be huge variability in the results. Only when you have tossed hundreds or even thousands of times do you begin to get an accurate estimate of a probability. A fair coin tossed twenty times might easily produce twelve heads, but that doesn't make the probability of a head 12/20 (or 60%). On the other hand, if a coin tossed 1000 times produces 600 heads, you have reasonable grounds to believe that the coin is biased.

3) **Collect some data.** What is the chance that the next person to walk past you will be wearing a hat? You can make an estimate by doing a survey: count the next hundred people who walk past. If five of them are wearing hats, 5/100, or 1/20, is a decent estimate of the probability that a random person on your street will be wearing a hat. Again, the larger the sample you take, the more reliable the estimate.

4) **Look at the past statistics.** What is the chance that your child's teacher has a birthday in September? Past statistics of birth dates will help you to come up with an estimate. Although September contains thirty of the 365 days in the year (just under one twelfth), past statistics suggest that slightly more than one twelfth of babies are born in that month (due to activity before the previous Christmas, perhaps?). That statistic, somewhere close to 10%, would form a good estimate of the chance of the teacher's birthday being in September.

5) **Use your 'expertise' (or a hunch).** Experts are asked all the time to comment on the chance of something happening. What chance Manchester United will win their next game? What chance the UK economy will drop into recession? Real 'odds' will be quoted for these things, and bookmakers rely on these odds being fairly reliable, but they are no more than educated guesses.

Randomness

Suppose you tossed a fair coin five times. To your surprise, you get five heads in a row:

H H H H H

What is the chance that the sixth toss of the coin will also be a head? Your logical side will be telling you that it has a 50-50 chance of being a head, but your emotional side will be saying '*surely* a tail is now due'. The tug of this emotional reaction can play havoc with our understanding of probability, because we begin to think of the coin as somehow knowing what it is doing.

But of course a coin has no 'memory', what happens next is completely unaffected by what happened before (the probabilities of each toss is described as being *independent*). A sequence of five heads in a row is just as likely as a sequence that goes, say, H T T H T (both will happen about 1 time in every 32 goes), but the sequence of heads grabs our attention.

The magician Derren Brown played on the public's emotional reaction to chance in one of his shows. Having established that a coin was fair by tossing it several times, he went on to show that he could 'control' the result by tossing ten heads in a row. What he then went on to show was that all they had done was to repeatedly film the same thing over and over again until he chanced upon getting all those heads.

Probably the most famous example happened in Monte Carlo in 1913, when the roulette wheel came up black twenty-six times in a row. As the number of consecutive blacks grew, gamblers began to massively increase their bets on red based on the fallacy that a run of reds was 'due' in order to balance things out. Gamblers bet far more money than they could afford and many lost fortunes.

An interesting experiment is to ask a teenager to pretend to be a coin, and to come up with a realistic sequence of thirty heads/tails. Their sequence might look like this: H H T H H T T H T T T H H T H T H H T T H T H . . . It looks realistic enough. But now toss a real coin and record its heads and tails. When the authors tossed a coin, this was the first sequence we recorded: H T T T T H T T H T H T T T T H T H T H H T T T H H H H H T . . . What's the difference? Real coin sequences will often have runs of four or five of the same side of the coin (notice the four tails near the start), whereas people's attempts to be 'realistic' tend to keep runs of heads or tails to a maximum of three.

Higher or lower?

One of television's more successful TV gameshows over the years has been *Play Your Cards Right*, originally hosted by Bruce Forsyth and since then by other presenters. It's still a popular game in pubs. The idea is simple: a playing card from a shuffled pack is turned over, and the contestant has to guess whether the next card will be higher or lower.

There are fifty-two cards in a pack, four suits, and in this game the cards (lowest to highest) are 2, 3, 4, 5, 6, 7, 8, 9, 10, Jack, Queen, King and Ace. The best card to draw is a two or an Ace – because you can be almost certain the next card will be higher (than the two) or lower (than the Ace).* Next best cards are the three and King, then the four and Queen, and so on. The very worst card is the eight, since it is right in the middle.

What do you do if you draw an eight? There are twenty-four cards that are higher and twenty-four cards that are lower, so you might as well guess. Unless, that is, you have already seen some of the other cards of the pack. One of the authors was in exactly this position in a pub game, with a jackpot of £200 on offer. The eight of Spades was turned over. A previous contestant had already revealed a seven and a two, which were still on display. This meant that the pack now had twenty-four cards higher than an eight and only twenty-two lower. The odds were therefore slightly in favour of calling 'higher'. The author did this – and ended up winning the jackpot. Maths does sometimes pay.

* You might be very unlucky and draw another two (or Ace).

Probability of More Than One Event

One area that teenagers often find difficult in probability is finding all the possible outcomes when there is more than one event. A simple example is finding the probability of getting two heads if you toss two coins. What are the possible outcomes? If you toss two coins then you will get either two heads, one head/one tail, or two tails. It seems there are three outcomes and so a probability of $\frac{1}{3}$ for two heads seems a good guess.

However, this is where you have to be so careful listing the outcomes. In this case, it makes things easier to follow if we label the two coins that are tossed, for example, let's suppose one is a 1p coin and the other a 2p coin. The possible outcomes are:

	1p coin	2p coin
1.	Heads	Heads
2.	Heads	Tails
3.	Tails	Heads
4.	Tails	Tails

Now we can see that there are four possible outcomes, because there are two different ways of getting one head and one tail. Only one of the four outcomes is two heads. So the chance of getting two heads is $\frac{1}{4}$. The probability of two tails is also $\frac{1}{4}$, while the probability of one head and one tail is $\frac{2}{4}$ (a head followed by a tail or a tail followed by a head).

One way of recording all the possible outcomes is to use a tree. Rather than using heads and tails, let's use another example. A bag contains five red and five black balls. You reach in and without looking you take out a ball and set it aside. You reach in again and take out a second ball and put it with one that you removed first. What is the probability that you have taken out two red balls?

The tree diagram for this problem is set up in the following way. The first ball removed will be either red or black and since there are equal numbers of both, the probability for each of these outcomes is 1/2.

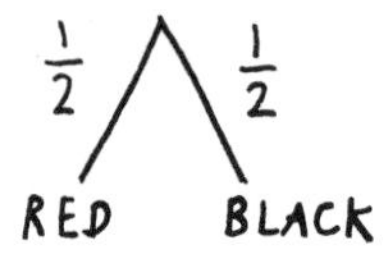

The second layer of branches is constructed on the end of each of these branches. Once again there are two outcomes at the second stage, red or black.

However, because the first ball was set aside (and not put back in the bag) the probabilities at this second stage are not fifty-fifty. If a red ball was removed first, then of the nine balls still in the bag there are now four red and five black, so the probability of pulling out a second red is 4/9 and a black 5/9. Similarly if a black was drawn out first, then the probability of the second ball being red is 5/9 and of being black is 4/9.

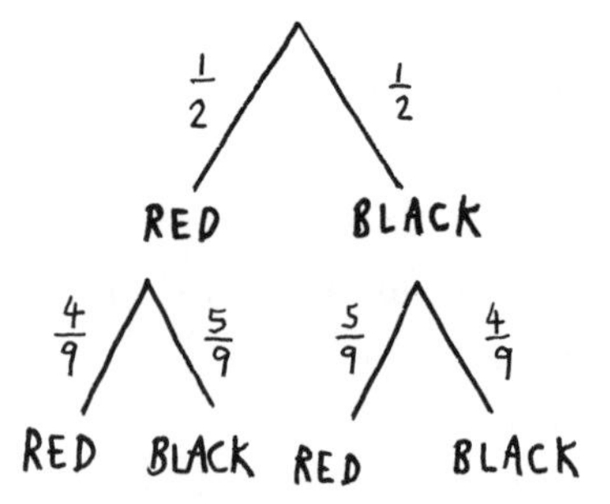

You can follow the branches of the tree to see every possible outcome. For example, there is a 1/2 chance that the first ball is red, and having gone down that branch, the chance that the second ball is red is 4/9. To work out the combined probability, think of this 'tree' as a set of paths down which you are dropping ball bearings. Half of the ball bearings will drop on the red side and half on the black side. Of the half that drop on the red side, 4/9 will drop down the lower red branch. So the chance of two reds is $\frac{1}{2} \times 4/9$, and half of 4/9 is 2/9. (Combined probabilities often involves multiplying two fractions together like this. Probability is one place where all that effort put into learning fraction arithmetic begins to pay off!)

The red/black balls also introduce another important feature of probability. If you toss a coin twice, the outcome of the second toss is not affected by the first toss: the two events are described as being *independent*. However, in the balls example, the chance of the second ball being a red depends on what ball you drew first. If the first ball was red, then the chance of a second red is 4/9, but if the first ball was black, the chance of the second being red is 5/9. When one event influences the probability of another, the two events are said to be 'dependent'. Understanding when probabilities are dependent and independent is important – and a huge stumbling block for many teenagers.

TEST YOURSELF

a) **Lucky red.** At a fairground stall, you get to win a prize if you toss a coin and get a head AND you then draw a red card (Heart or Diamond) from a normal pack of cards. What is the chance that you will win?

b) **Weather forecast.** This is a true story. In a television weather forecast, the forecaster announced that there was a 50% chance it was going to rain on Saturday and a 50% chance it would rain on Sunday. 'This means,' he said, 'that the chance it will rain at some point over the weekend is . . .'. What was the correct answer?

Two dice games

Many tactical games involve probability. One of the best things about such games is that it is possible to enjoy (and even win) these games without fully understanding the maths behind them – but knowing the maths certainly helps.

1) **Horse race.** There are twelve horses in a race, numbered 1 to 12. The organiser of this race rolls two dice and adds up the scores. The horse whose number is that total moves forward one space (so if the organiser rolls a 3 and a 6, say, scoring 9, then horse 9 moves forward one space). Which horse would you bet on to get to the finish line first?

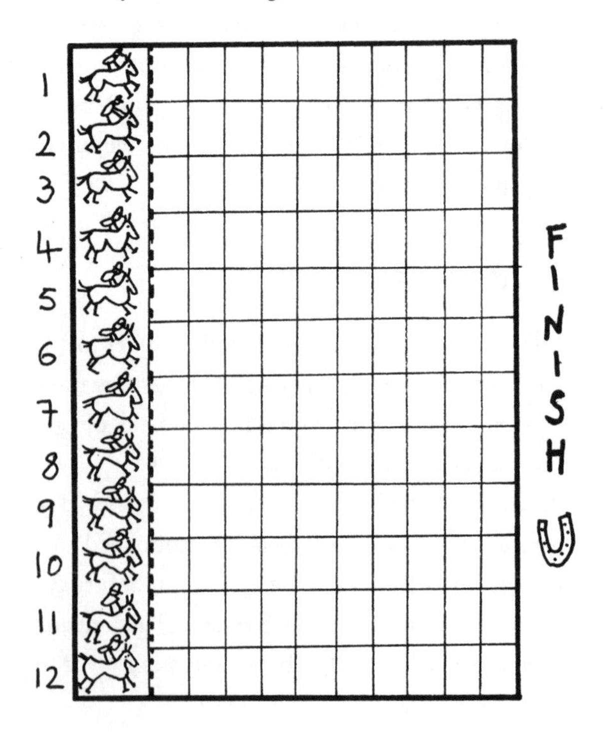

2) **Wipeout.** You roll a dice, and keep adding the score. However, if you roll a 6 you lose all your points. You can 'stick' on any score you want. Your aim is to get as big a score as possible. Should you stick if you get to a total of 10? Or should you keep going? When is it sensible to stop?

The solutions to these dice game problems are on pages 347–9.

PROBABILITY
If you do only three things . . .

- Make sure your teenager is clear about what the probability of a particular outcome means. Just because there are two outcomes doesn't mean that the chance of each outcome is 50:50. To take an extreme example, there are two possible outcomes when you go on a bike ride: either you will fall off the bike or you won't. This doesn't mean there is a 50:50 you will fall off. It's helpful to think of probability as 'If these circumstances (e.g., going for a bike ride) were to happen one hundred times, how many times would I expect the particular outcome (e.g., falling off) to happen?'
- Be prepared to find the probability of some outcomes extremely surprising. Small changes in the way you state a situation can have a huge effect on its chances. For example, if you randomly pick two people out of a group of twenty, then the chance that those two people will have the same birthday is 1 in 365 (the number of days in the year). But the chance that within that group of twenty there exist two people who share a birthday is close to 50%. This is partly explained by the fact that within a group of twenty, there are about 200 different pairs of people, which massively increases the chance that at least one of those pairs will have shared birthdays – but it still a very surprising result.
- Few teenagers (or adults) are convinced by the maths that predicts probability. When possible, encourage your

teenager to investigate probability by doing repeated experiments and recording the results. For example, if you pick a group of twenty-two footballers from all the Premier League matches one weekend and check their birthdays, you will find that in about half the matches there are at least two players with the same birthday, confirming the outcome we mentioned earlier.

PRESENTING AND INTERPRETING STATISTICS

Your teenager will deal with statistics in three ways, often referred to as the 'data-handling cycle' because the three parts follow on from each other in a logical cycle:

1) Collecting data.
2) Presenting that data in a useful form.
3) Interpreting the results to come up with useful conclusions.

Traditionally, it was the 'presentation' part that was emphasised in school mathematics, particularly the time spent drawing up graphs and working out averages. Thanks to computer technology, however, the presentation can now be done at the click of a mouse, so there's more time spent today on the interpretation of those statistics.

The basics of statistics haven't changed in decades, but some of the techniques used in school certainly have. In particular, you are likely to come across some new terminology: 'Box and whiskers' and 'Stem and leaf' are two formats for presenting statistics that your teenager will need to learn. We explain these later.

Three Types of Average: Mode, Median and Mean

If you want to be able to sum up a sportsman's performance or the lifestyle of a typical member of the public, then nothing beats an average. It's a way of conveying a lot of information in a single number.

The history of the word 'average' is not mathematical at all. It originally referred to goods that were damaged at sea and lost cargo. Then it evolved to mean sharing the costs of those losses between the different owners of the cargo, and from there it became a general term for spreading irregular numbers more evenly. Today, it is best summed up as 'typical'.

When the word average appears on its own, it usually refers one of the following three different measures:

The mean

By far the most common type of average is what is known as the *mean*. To find it, add up all the values and divide the total by the number. For example, if a long jumper's performances so far this year have been (in metres) the following six jumps: 6.93, 6.85, 7.07, 7.26, 6.99 and 7.38 then his mean score is worked out by adding all those jumps together (that's 42.48 metres, to save you working it out), and dividing by 6. His average (mean) jump distance has been 7.08 metres. If his average last year was 6.99 metres, this is a sign that he's in form, and maybe it's a trend.

If you hear the word 'average' being used in a news story, there is a high chance that it refers to a mean.

The median

The main rival for the mean is called the *median*. This is the middle value in a set of data, and to find it, you have to sort all the data into numerical order – very easy to do with a spreadsheet. In the case of our long jumper, his distances in ascending order were 6.85, 6.93, 6.99, 7.07, 7.26 and 7.38. Since there are six numbers here, the 'middle' one is actually between the third (6.99) and the fourth (7.07). The normal thing to do here is simply find the midpoint of those two numbers, so the median distance of the long jumper is 7.03 metres.

You'll see that the mean (7.08) and the median (7.03) are very close to each other in this example. Indeed very often these two averages are close to each other, which is why people tend to get sloppy and not bother saying which type of average is being referred to.

However, means and medians can be very different if the data you are looking at is skewed – that is, if when plotted as a graph it is not symmetrical (see the income example below).

The mode

The third type of average is called the *mode*. This picks out the value that occurs most frequently, for example the most popular shoe size. It's therefore a good indicator of the most 'typical' result. A men's shoe shop might sell shoes in sizes from 6 to 13, but may sell more size 9s than any other size (the mode is 9) – it would make sense to order in more of that size shoe than any other.

The mode can also be helpful with data that doesn't involve numbers. For example, you could categorise the UK's population according to hair colour, and find that brown is the most frequent. If you hear a claim that the 'average Brit' has brown hair, it must be the mode that is being referred to (the mean and median would be meaningless in this case).

With data that is spread over a wide range, such as the long jumper's results, the mode can also be meaningless, because no values appear more than once. One way to get around this is by grouping the results into bands (for the long jumper, these could be 6.80–6.99, 7.00–7.19, 7.20–7.39). The mode of the jumps then turns out to be in the range 6.80–6.99, suggesting that the jumper's 'typical' jump is in this range, even though his mean and median are higher.

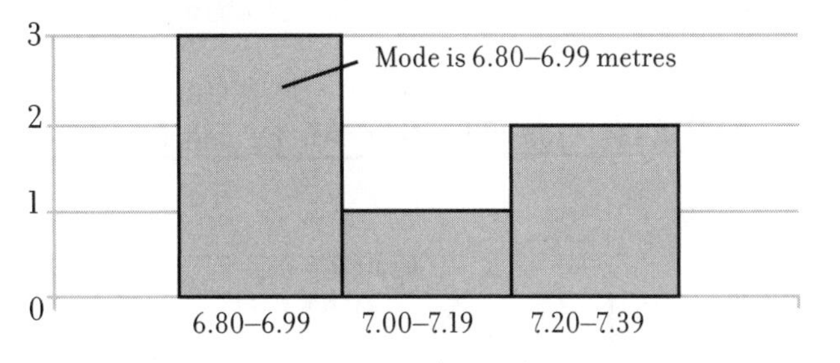

Type of 'average'	What it is	Advantages	Disadvantages
Mean	Add all the numbers and divide by how many numbers there are.	Quick and easy, especially with a calculator or spreadsheet. The most commonly used, and understood, average.	It can be misleading when used for data that is 'skewed' (such as personal incomes).
Median	The middle value when all the numbers are set out in ascending order.	It really is midway, half of the population is above it and half is below it, so gives a 'fair' impression.	Requires more sorting to work it out than the mean.

Mode	The number or property that has arisen most often.	Instantly sums up what the most common occurrences are.	Not very useful for data that has a huge spread, and it's possible for a set of data to have at least two different modes.

Why most people's income is 'below average'

Few things cause more social anguish than statistics about 'average' incomes. Here's a graph of the incomes of adults in the UK from a few years ago (the data hasn't changed much since then).

DISTRIBUTION OF INCOME IN UK (2010)

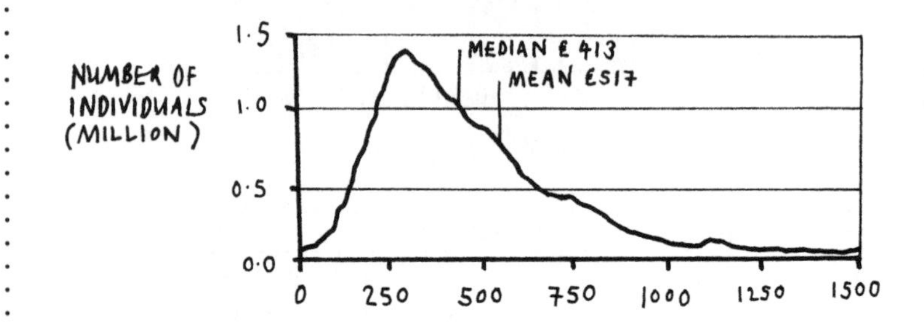

Notice how most people seem to be clustered in the range £200–£500 per week, with the peak (modal) income around the £300 mark. However, there's a small but significant number who earn far more, including some, such as footballers and financiers, who are earning tens of thousands of pounds every week. The result is a graph that is not symmetrical. The peak occurs towards the left, and there is a long tail stretching to the right. This is known as a *skewed* distribution.

The median value is just over £400: half of all adults earned more than this and half earned less. However, it is often not the median but the mean income that is quoted in stories about average income. The mean value is considerably higher, over £500. This is because the minority with huge salaries pull up the value of the mean, whereas the median would stay where it is even if those very wealthy people's incomes were to double. So when it comes to income, it's quite possible that about 75% of adults earn less than the 'average' (mean) income – which gives politicians and union leaders plenty of opportunity for agitation.

The 'Spread' of Data

Although the average of a set of data is an important guide, it can be just as important to know about the spread of data – as the examples of the long jumper and the UK incomes showed.

The easiest guide to the spread is simply to state the highest and lowest values. For example, suppose a group of students get the following scores on a test: 73, 42, 66, 78, 99, 84, 91, 82, 87, 94, 45, 56.

A first step in processing the data is to put them in order: 42, 45, 56, 66, 73, 78, 82, 84, 87, 91, 94, 99

Now that the data is in order, the 'range' is easily found – it's the difference between the lowest and highest scores so in this case the range is 99–42 or 57 points. On its own the range may not tell us much but if these scores were compared to other groups' scores on the test, the range would give an indication of whether this set of results was typical or not.

Stem and Leaf

There are lots of ways in which you can present the data to give a visual impression of its spread. Most parents will have plotted bar charts similar to the one on page 293 but a nice 'back of envelope' technique called 'stem and leaf' is often taught these days. It has the advantage that you can just scribble it down, instead of having to plot it.

The data is presented in two columns, with the 'stem' as the left-hand column and the 'leaves' listed in the right-hand column. In most examples that your teenager will meet, the stem will be a column of numbers representing the first digits of the data (the 'tens' digit). The leaves will be the units of each number. As an example, let's do a steam-and-leaf diagram for the set of test scores on the previous page. The first two numbers in the ordered list of test scores is 42 and 45. These both have a stem of 4, and the 2 and 5 are recorded alongside. The whole table looks like this:

TEST RESULTS

Stem (tens)	*Leaf (units)*
4	2, 5
5	6
6	6
7	3, 8
8	2, 4, 7
9	1, 4, 9

Stem-and-leaf diagrams give a quick impression of how the data is spread, and don't need any careful plotting.

The table helps to pick up a pattern. It looks like the scores on this test are 'skewed' a bit towards the upper end (in the 80s and 90s).

TEST YOURSELF

At a baby clinic, a group of ten-week-old babies are weighed. Their weights in kilograms are: 5.2, 4.8, 6.7, 6.2, 5.4, 6.2, 5.0, 5.1, 5.5, 5.8, 4.7, 5.6.

a) Find the mean and median weights.
b) Set these out in a stem-and-leaf table.
c) What weight is the mode?

Interquartile Range

In most sets of data, the values at the extremes might be extremely untypical. For example, in the school test results, what if one student completely flunked the exam and scored zero. The range of scores would now be 0 to 99, but that suggests a much wider spread of typical scores than was actually the case. Another statistic that helps is the *interquartile range*. As its name suggests, this is based on dividing the data set into quarters, and this range gives the 'spread' of the middle 50% of the data.

With our set of twelve test scores, the cut-off for the bottom 25% lies between the 3rd and 4th values, and for the top 25% between the 9th and 10th. As with the median, we take the middle of each of these pairs. The lower interquartile point is $(56 + 66) \div 2 = 61$ and the upper interquartile point is $(87 + 91) \div 2 = 89$. So the interquartile range is the difference between the higher and lower quartile, $89 - 61 = 28$. The graph shows where these points are:

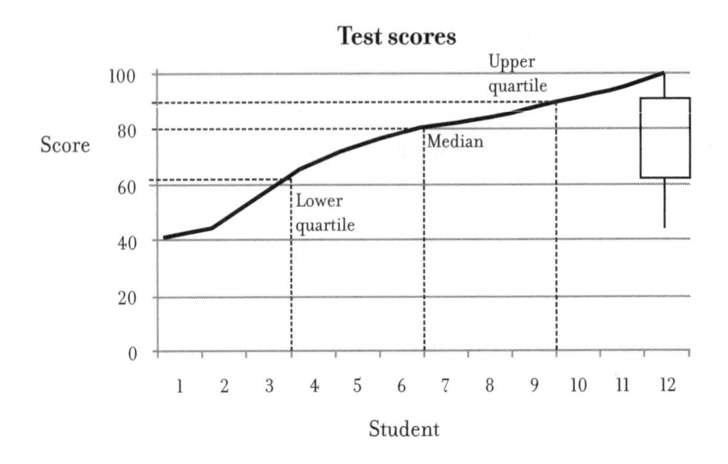

So what does all this tell us? The overall range of the marks is 57 but the interquartile range is 27 – this latter figure tells us that 50% of the scores have a range of 27 points, so this middle set of scores is a little bit more 'bunched up', with the scores ranging from 61 to 89. A new learner joining the class, taking the test and getting a score of 60 might be quite pleased with that, but this simple analysis tells us that a score of 60% is in the bottom 25% of the class, so not that great (indeed, even a score of 78 is in the bottom 50%, as it is below the median score of 80).

Incidentally, quartiles are related to 'percentiles', a word familiar to any parent who had their baby's weight and height monitored for the first few months of life. If you are told that your baby is at the 75th percentile of weight, this is the same as the upper quartile – 25% of babies are likely to be heavier, and 75% lighter. If you look at a baby's percentile graph, you'll notice that the 25th and 75th percentile lines are closer to each other than they are to the 5th and 95th percentiles – typical of the way data tends to be concentrated towards the middle.

Box and Whiskers

Another way of presenting a range of data is known as the Box and Whiskers plot, a diagram only invented in the late 1970s (which explains why many parents have never heard of it). Instead of a graph, the data is condensed into a simple scale diagram. In the centre of the diagram is a rectangular box, whose ends represent the upper and lower quartiles, with a line across it to show the median. Out of the box come two lines (whiskers), which usually extend to the highest and lowest values, indicating the range.

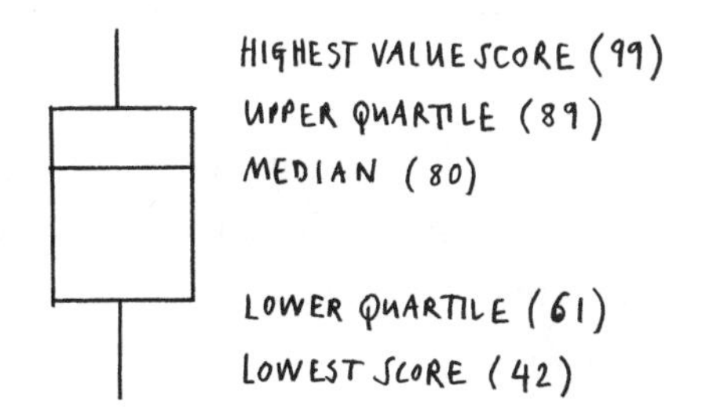

If you look back at the graph of scores, we've superimposed the Box and Whiskers plot on the right-hand side so you can see how the two ways of presenting the data compare. As the graph of the scores over 80 'flattens out' so the top half of the box gets 'squashed'. The attraction of Box and Whisker diagrams is that they are quicker to draw than graphs but still give a good impression of the 'shape' of the data.

The Normal Distribution

If you roll four dice dozens of times and record the total scores, or examine the heights of a class of fourteen-year-olds, or look up the peak temperatures in March for the last thirty years, you'll find that these apparently unrelated statistics all have a connection. If you plot how often each result comes up, the shape of the graph (the *frequency* graph) will almost certainly have a domed or bell shape, like this:

The more data you collect, the smoother the curve will be. This is known as a *Normal distribution*, and what is remarkable is that the curve's formula applies just about everywhere, from natural phenomena to economics – in fact, anywhere that statistics have a fluctuating, random element.

This normal curve is extremely useful in statistics as it can be used to make all sorts of predictions, such as the number of defect items that will come off a factory's production line, or the number of people that will require size-13 shoes.

It's unlikely that your teenager will do any formal maths with the Normal curve up to GCSE, but they'll certainly encounter the shape.

Pie Charts and Bar Charts

Two common forms of presenting data are pie charts and bar charts. The reason for choosing one over the other depends on what it is you want to emphasise.

A poll in the USA surveyed teenagers on their favourite leisure activities and came up with the following results:

Activity	*2004*	*2007*
Reading	35%	29%
TV	21%	19%
Time with friends and family	20%	14%
Computer	7%	8%
Movie going	10%	7%
Other	7%	23%

On a bar chart the results look like this:

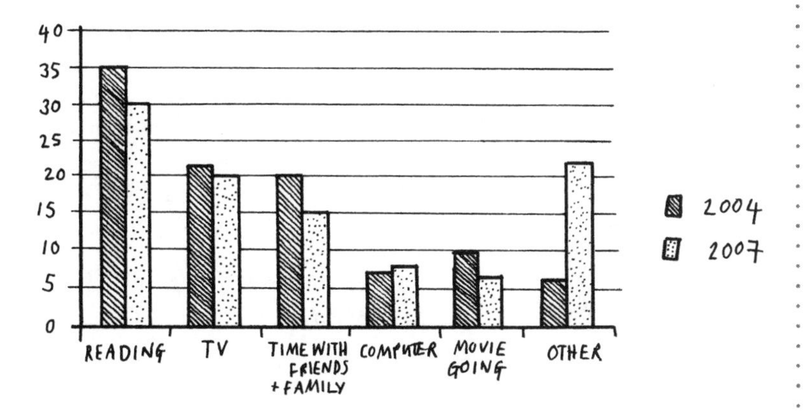

What the bar chart reveals is how teenage interests changed between the two years. We can see that reading, TV and time with friends and family were still the most popular activities in 2007 but each had gone down in comparison with 2004 (the gains in 'other' turning out to be sport and art/craft activities).

Presented as two pie charts the data looks like this:

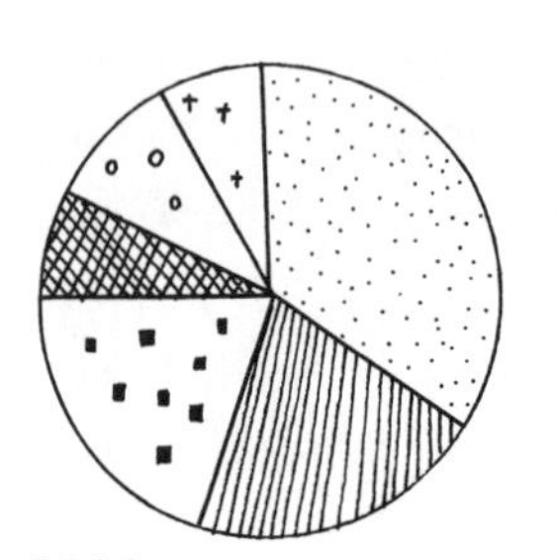

2004

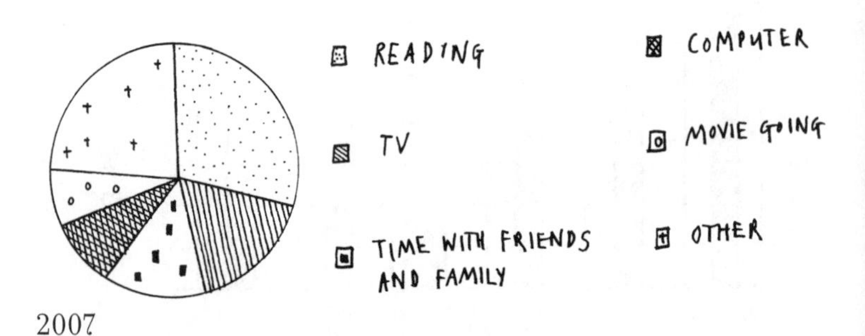

2007

It is less easy to compare across the years in the pie charts – how much different is that 'slice' of reading in 2007 compared to 2004? But now we can see more about what is going on within each year. For example, we can see that taken together reading and TV were chosen by over half of the teenagers in 2004 but this accounts for less than half in 2007.

Pie charts – a common teenage error

A common error that teenagers make with pie charts is confusing the proportions with the absolute numbers. Imagine the two charts were actually comparing the preferences of teenagers from two different schools (School 2004 and School 2007). We can say that a greater proportion of the students at School 2004 expressed a preference for reading compared to the proportion expressing that preference at School 2007. But we can't say that more teenagers are reading at School 2004 than at School 2007 as we do not know how many attend each school. Maybe there were 1000 students at School 2007 (of whom 290 loved reading) and only 500 at School 2004 (of whom 175 loved reading).

Histograms

There's another form of diagram known as a histogram. Histograms are similar to bar charts, but while the width of each column in a bar chart is the same, in histograms the bars can be different widths, to represent the range of data that they are covering.

For example, here's a histogram of the weights of babies from page 289 (rearranged in order of increasing weight they are: 4.7, 4.8, 5.0, 5.1, 5.2, 5.4, 5.5, 5.6, 5.8, 6.2, 6.2, 6.7):

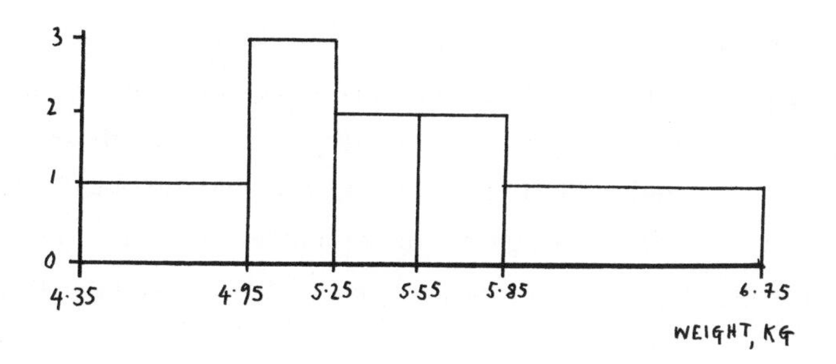

Notice how the bars at the end are wider than the three bars in the centre. This is because they represent a wider range of data. The three bars in the middle represent a spread of 0.3kg; there were three babies in the range 4.95–5.25kg, and two in the ranges 5.25–5.55kg and 5.55–5.85kg.

Meanwhile, two babies were in the range 4.35–4.95kg, but because this range is twice as wide, the height of the bar is halved. For the same reason the height of the bar in the range 5.85–6.75kg, representing three babies, is divided by 3. This means that the area of each bar accurately represents how many babies were in that region of the graph.

It's a clever idea, because it means the histogram gives a good impression of the distribution of the data, with the bell shape associated with a Normal distribution (page 292). However, it also causes teenagers considerable grief, as they regularly fail to spot that the ranges of data in a table are not equal.

Incidentally, histograms were invented at the end of the nineteenth century by an English mathematician called Karl Pearson. 'Histo-' comes from the Greek word for sail: a histogram (sort of) resembles a ship sailing past with various different sails raised.

Correlation and Causation

Often there is a connection between two sets of data. For example, if you gathered a group of people together and measured their foot size and their height, the chances are that those with large feet would tend to be taller. These statistics are said to be *correlated.* If the value of one thing tends to increase when the other increases, this is called *positive* correlation and if one thing tends to decrease when the other decreases it is called *negative* or *inverse* correlation. An example of negative correlation would be air temperature plotted against altitude. The further up a mountain that you climb, the colder the air gets.

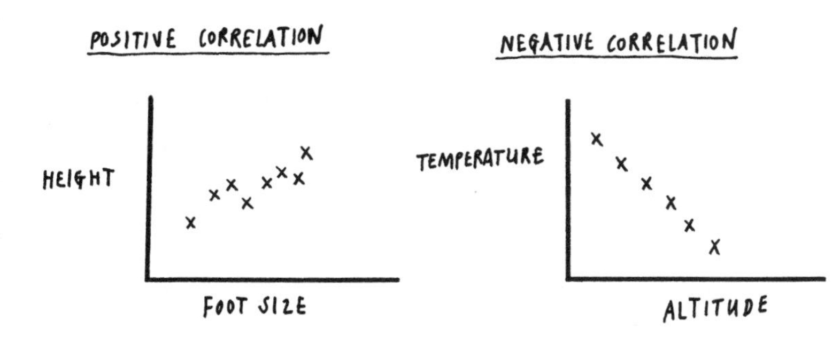

Notice how in these examples the points on the graph of height versus foot size don't form a smooth line, whereas those in the temperature/altitude graph do. The straighter the line, the *stronger* the (linear) correlation is between the two variables.

Correlation is one of the most useful things that statistics can reveal. It can help to identify health risks (such as the correlation between smoking and lung disease) and it can highlight surprising connections that weren't expected, such as the fact that having a picture of a baby in your wallet significantly increases the chance that the wallet will be returned to you if you lose it. It's an important concept for any teenager to understand, especially if they want to go on to study science.

However, there is an even more important thing to understand about correlation, which is that humans can read too much into it. We tend to assume that because two variables are correlated, it must mean that one has caused the other.

Take these examples:

1) The more graffiti there is, the higher the levels of crime in the area. Conclusion? Maybe graffiti causes crime, by creating a 'criminal atmosphere'.
2) A runny nose tends to go with a sore throat, so runny noses cause sore throats.
3) The number of drownings tends to increase roughly in line with ice-cream sales, so ice creams cause drownings.
4) Reading ability is linked to shoe size, therefore bigger shoes improve your ability to read.

In each of these examples there is an alternative interpretation. Sometimes the cause might actually be the other way round: for example, it could be that greater levels of crime mean there are more people out there wanting to write graffiti.

However, just as often, correlation is due to some third factor that is affecting both of the variables. Graffiti and crime might both be caused by economic and social deprivation. A runny nose and sore throat are both caused by a cold, they are both symptoms, not causes. Ice-cream sales and drownings both increase during hot weather (when more people tend to go out swimming). And as for reading ability and shoe size: small shoes are worn by young children, whose reading ability is of course lower than that of adults. The causal factor here is age!

Understanding the difference between correlation and *causation* is extremely important. As an adult, it helps you to make informed decisions about things like the alleged link (now almost entirely discredited by scientists) between the MMR vaccine and autism. Most of the maths behind it is of A level standard or beyond, but teenagers are expected to appreciate the principle without needing the higher maths.

Gathering Data: Surveys and Sample Size

All of the statistics that get interpreted have to come from somewhere, and these days teenagers are rightly taught about the methods and pitfalls of collecting data.

Much of the art of gathering data is down to psychology and common sense rather than maths. If you write a questionnaire with leading questions such as 'On a scale of 1 to 10, where 10 means the worst possible, how bad do you think the current government's education policy is?' then don't be surprised that you get very biased data as a result. The question is inviting a negative reaction, and regardless of the quality of the government policy, a large number of 10s would be likely whichever party was in power.

There is, however, also some serious maths that goes into the collection of data. Perhaps most important is being able to answer the question, '*How much data do I actually need to collect?*'

In the chapter on probability, we included an example of a bag containing a mixed set of ten red and black balls (page 275). Imagine that you don't know the contents of the bag, all you know is that it contains ten balls that could be any combination of reds and blacks. What you want to know is the proportion of each. You could of course simply take out all of the balls, but is that necessary? You can get a very good idea of the mix just by taking out a few balls.

Clearly one ball isn't enough: if it's red, you will be under the mistaken impression that all the balls in the bag are red. Two balls would be better, but there's roughly a 50% chance that the two balls will be the same colour, again giving you a misleading impression of the contents. The more balls you take out, the more likely it becomes that the split of balls that you have chosen is representative of the make-up of the entire bag. The question 'How many is enough?' depends on how confident you need to be. If you're happy to be right within a margin of error of 20%, you need a much smaller sample than if you need to be 99% accurate. To be absolutely certain you'd need to remove all ten balls from the bag.

Any kind of survey, whether it's asking shops to report on the number of TVs they have sold or asking people who they intend to vote for, is rather like putting your hand into a bag and drawing out a coloured ball. And the more diverse the population you are dealing with, the bigger the sample you have to take – it's harder to work out the composition of a bag containing several different-coloured balls than one containing only two colours.

Data gathering and probability are much more closely linked than you might think.

TEST YOURSELF

Suppose you are conducting a random survey to get an 'accurate' estimate of the proportion of shoppers who have watched football on TV in the last month, and also the proportion of British people that have a pet snake. What is your hunch for the number of people you might have to interview to get a good result in each of these two surveys? Why?

Unreliable confessions

Every year the NHS conducts a survey on the health and lifestyle of the nation. One of its more recent studies included a question about the sexual behaviour of the respondents. The results were revealing: women aged sixteen to sixty-nine reported that on average they had had 4.7 sexual relationships with members of the opposite sex so far in their life. Meanwhile, men reported an average of 9.3 sexual relationships with members of the opposite sex, almost double the figure for women. Your first conclusion from this might be to think that men are more promiscuous than women – until you remember that fundamental law of nature that it takes one man and one woman to make an opposite-sex relationship. Since the number of men and women is the same, the average (mean) number of opposite-gender partners for men and women should both be the same.

What is going on? It could be that the 'average' being referred to was not a mean but a median. This would allow for the majority of women having fewer partners than men (lowering the median for women), while a small number of women might have a very large number of male partners to make up the numbers. Or it could be that people outside the sample age group of sixteen to sixty-nine-year-olds were involved.

There is, however, another perhaps more plausible explanation. It might well be the case that either the men or the women (maybe both) are not reporting their experiences accurately. After all, memory can play strange tricks. Especially when you have your reputation to think about.

Lies, Damn Lies and Statistics

There is a wonderful book called *How to Lie with Statistics*, written back in 1954 yet still in print, and many of its messages are as true today as they ever were. The presentation of statistics can be manipulated to convey extremely misleading messages without actually 'lying'. Your teenager will learn about some of the common tricks that are used, and you can help by looking for them in news stories and advertisements.

Here are three of the most common ploys:

Vertical axis doesn't start at zero

This (fictitious) graph shows the decline in the number of a particular species of bird over five years:

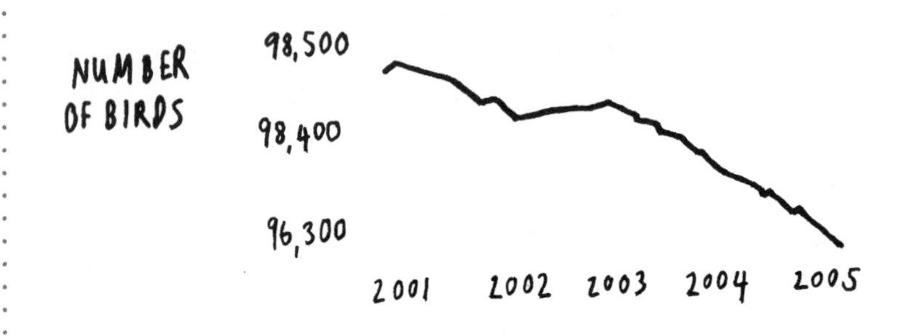

It looks alarming at first glance, until you notice that the actual drop has been about 200 birds out of 98,000, a fall of about 0.2%.

Misleading pictograms

The chart indicates that the number of people working in a factory has doubled in the last year.

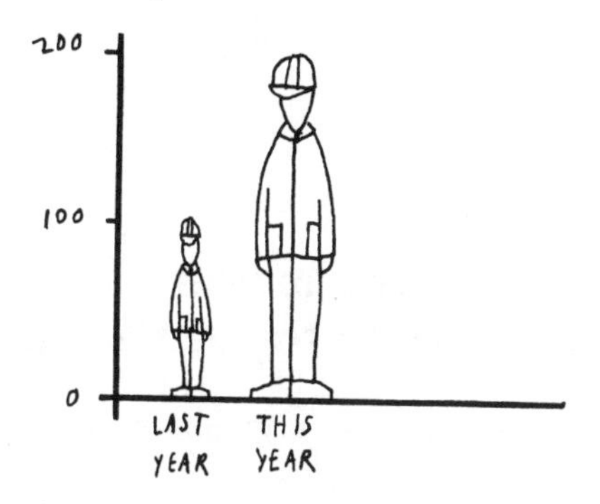

The height of the character has doubled; however, the impression given is that numbers have quadrupled, because his width has also doubled, making him four times as large.

Different scales

Here's a chart showing how home energy bills in one country varied in comparison with the price of oil. In 2008, while oil prices had risen by around 50%, household bills had gone up by a

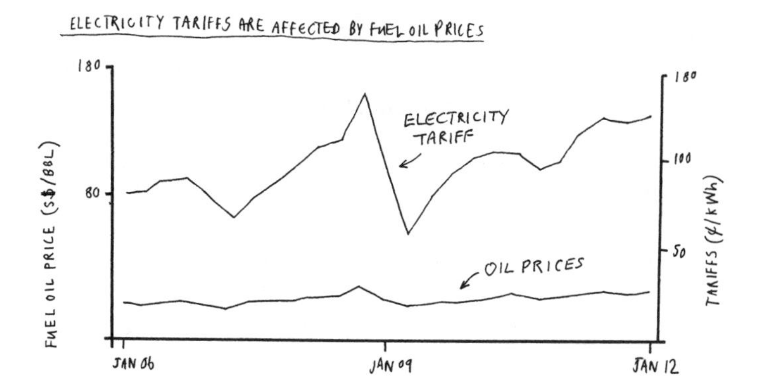

shocking 90%. But whoever produced the chart decided to make it look even more dramatic by using different scales. Oil prices are squashed down at the bottom so that even big variations look small, while the variation in household bills has been magnified by using the whole scale.

STATISTICS
If you do only three things . . .

- Keep an eye out for charts and graphs in advertisements or politically motivated stories that you can chat about with your teenager. In what ways are they manipulating the truth to create a particular message?
- Spreadsheets are a great way of generating lots of different graphical representations quickly – encourage your teenager to enter some data that interests them (statistics on allowances given to teenagers, say). If they wanted to make a case from the data (for example, to demonstrate that their weekly allowance should be higher) which representations are best?
- Collect reports of surveys from the news. Discuss with your teenager what you are told about who was involved in the survey and how this might affect the results.

PART FOUR

THE QUESTIONS & ANSWERS

QUESTIONS THAT YOUR TEENAGER MIGHT ENCOUNTER

Are you ready for a maths test? This section contains a collection of maths questions that a teenager might be confronted with. It's a chance to put yourself in their shoes and to remember what it was like to sit in an exam room, as the teacher announced, 'You may now turn over your paper.'

These questions don't cover the whole range of the curriculum (which keeps changing in any case), but they do at least give a flavour. There is a wide range of difficulty, although nearly all of the questions could have featured in a GCSE exam, many of them in a 'Higher Tier' paper. You might find some of the questions surprisingly difficult (wasn't GCSE supposed to be 'dumbed down'?). If so, you are not alone, many parents feel this way. On the other hand there are some parents who are confident with maths and who will breeze through these questions wondering what all the fuss is about.

When you've had a go, take a look at the answers and find out what other adults and teenagers make of these questions.

CALCULATORS

Exams are usually split into 'Calculator' and 'Non-calculator' papers, though having a calculator is often of little help in answering questions in papers where it is allowed. A calculator would probably be permitted for all except the first five questions. The only questions where a calculator is essential are numbers 6 and 20.

MARKS

In exam papers there is usually an indicator of how many marks are available for answering each part of the question. This helps to guide teenagers on where they need to concentrate their efforts, but it also provides unintended clues about what maths is involved in the answer. If a question is offering 4 marks and you work out an answer instantly, it almost certainly means that you've used the wrong maths. Likewise, questions offering 1 mark shouldn't require a page of workings. However, because our test isn't a formal exam, we've decided not to try to allocate marks for each segment.

CALCULATOR NOT ALLOWED FOR THE FIRST FIVE QUESTIONS

Q1 a) The price of a particular brand of baked of beans has risen from 60p to 69p. By what percentage has the price has increased?

b) Amy sees a dress she likes, which is marked as '20% off' and has a label with its new price of £40. What's the saving on the original price of the dress?

Q2 Work out $4 \times 10^5 \times 6 \times 10^{-2}$
and write the answer in standard form.

Q3 You are told that $7.8 \times 44 = 343.2$

Work out: a) 780×4.4

b) 3.9×8.8

Q4 The two triangles below are similar. What is the length L ?

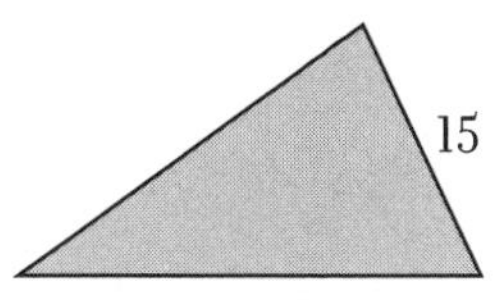

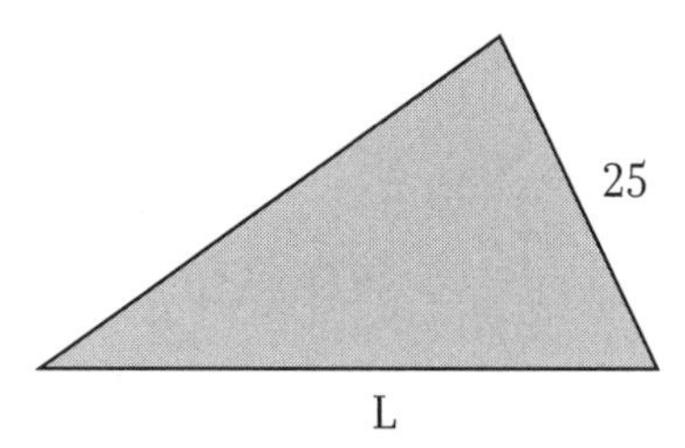

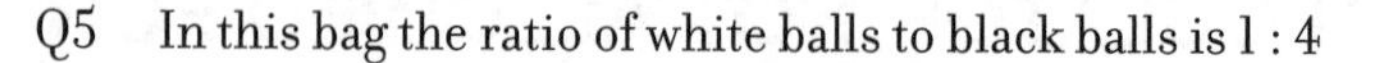

Q5 In this bag the ratio of white balls to black balls is 1 : 4

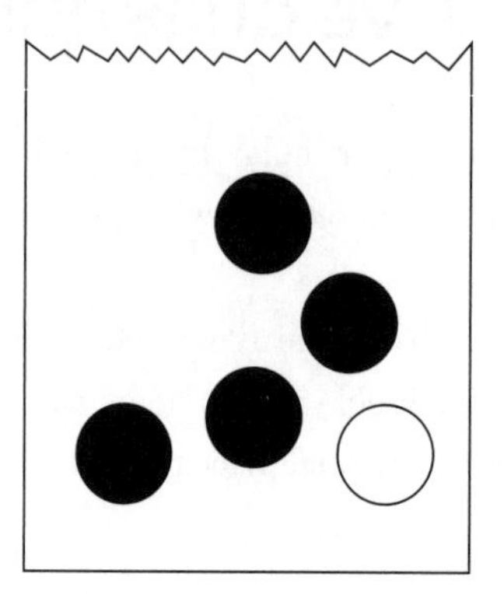

How many **more** white balls do you need to add to make the ratio of white to black 4 : 1?

Q6 In the triangle below, AB is 4 metres. Work out the length of AC:

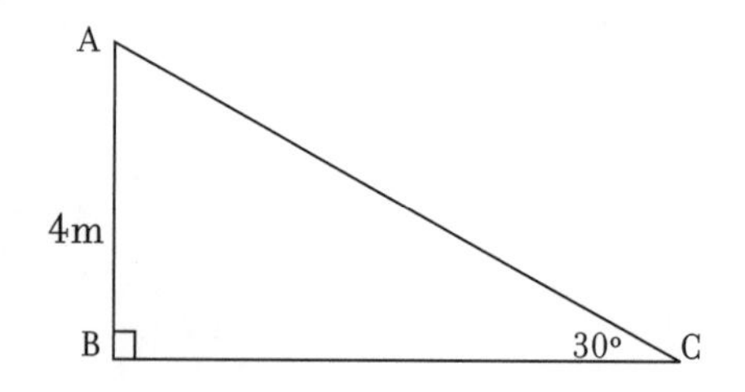

Q7 Look at these equations

$$4x + 3y = 32$$
$$30x - 30y = 30$$

Use both equations to work out the value of y.

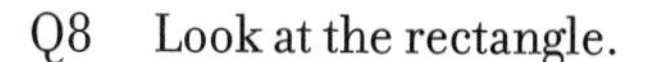

Q8 Look at the rectangle.

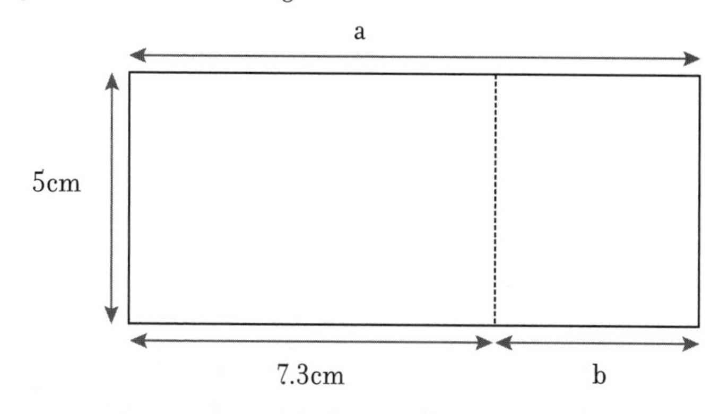

The total area of the rectangle is $50\ \text{cm}^2$
Work out the lengths *a* and *b*.

Q9 If $\dfrac{2(y+1)}{7} = 4$, what is y?

Q10

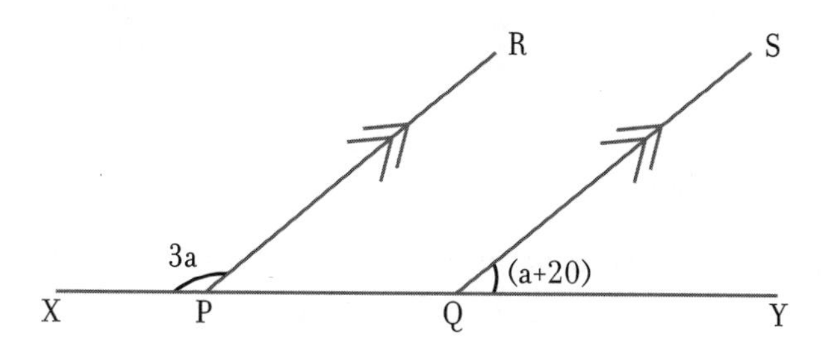

XPQY is a straight line. Lines PR and QS are parallel to each other.
Angle XPR $= 3a$ degrees
Angle SQY $= (a + 20)$ degrees
Work out the value of a.

Q11 A drawer contains 20 individual socks, as follows:

6 black
4 white
4 red
3 blue
3 yellow

Ellie picks a sock from the drawer at random.

a) What is the probability that the sock will be white?

b) Having taken out a white sock, what is the probability that the next sock Ellie takes out will be a different colour?

Q12 Boxes A and B contain some marbles. Box A holds $5x + 2$ marbles while box B contains $3x + 14$ marbles.

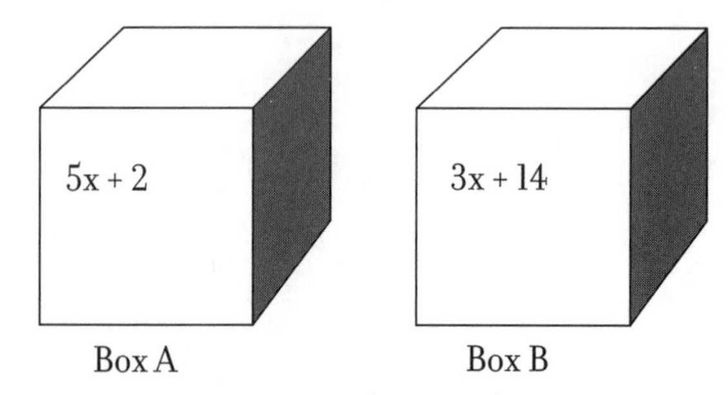

Box A contains MORE marbles than box B. What is the minimum number of marbles that box A must contain?

Q13 a) Expand and simplify: $(3x + 2)(4x - 1) - 5x$.

b) Factorise: $3x^2 + 7x + 2$

c) Use the answer to (b) to find a value of x for which $3x^2 + 7x + 2 = 0$

Q14

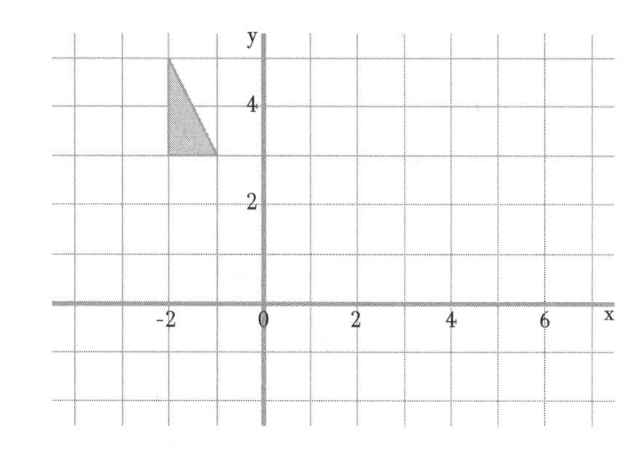

(a) On the diagram, draw the line that represents $y = x + 2$

(b) Draw the triangle when it has been reflected in the line $y = x + 2$

(c) Rotate your reflected triangle from (b) by 90° anti-clockwise around the point (0,2) and draw the triangle in its new position.

(d) What single transformation would take the triangle from its position in the diagram above to its final position after part (c)?

Q15 Rose has come up with a theory. She reckons that if n represents a prime number, then $2n + 1$ will also be a prime number.

Prove that she is wrong by showing a counter-example.

Q16 The line PQ is the diameter of a circle centre O.

R is a point on the circumference of the circle. The angle QPR is 30º.

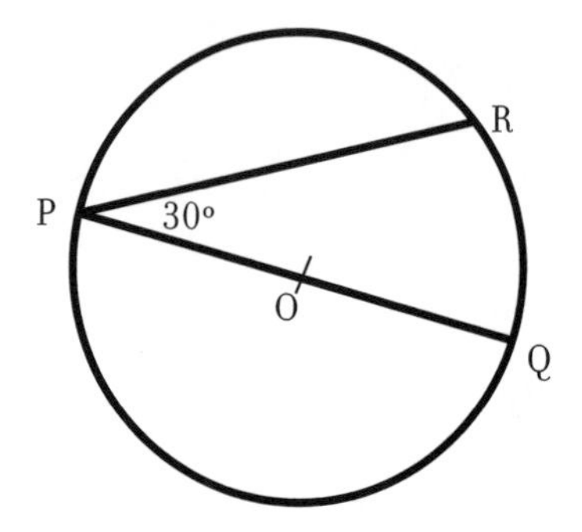

Draw the lines OR and RQ and without measuring any angles explain why:

a) The triangle formed by the points OPR is isosceles

b) Triangle ORQ is equilateral

Q17 Here is a sketch of the graph showing the amount of petrol in the tank of Sam's car over a period of six hours:

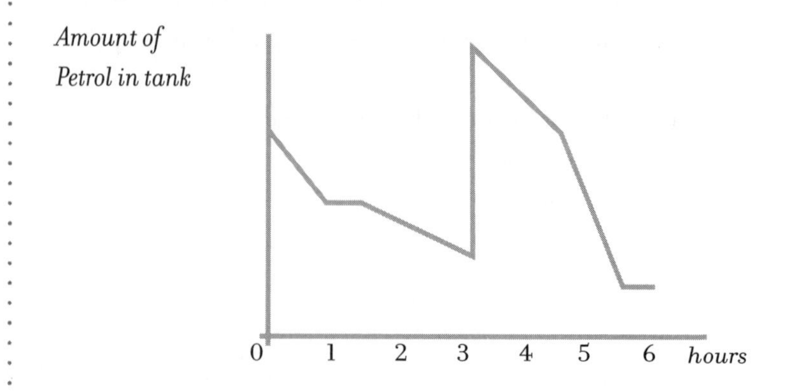

a) What happened after three hours?

b) For roughly how long would you say the car was stationary over the six hours?

c) Give a likely reason why the slope of the graph gets steeper after about 4 ½ hours.

Q18 Ali is thinking of two whole numbers, A and B.
A and B have a negative sum and a negative product.

Which statement or statements from the following could be true?

a) A and B are both negative

b) If you add 5 to A you get B

c) If you multiply A by 2 you get B.

d) $A^2 = B^2$

Q19 Jo has two fair four-sided 'tetrahedron' dice. The dice are both numbered 2, 5, 7 and 10.

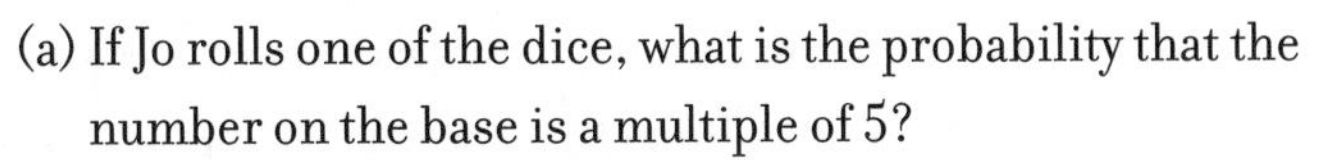

(a) If Jo rolls one of the dice, what is the probability that the number on the base is a multiple of 5?

(b) Jo is going to roll both dice and multiply the scores of the numbers on the bases together. What is the probability that the product will be a multiple of 5?

Q20 The number of people diagnosed with a serious strain of flu during the first few weeks of an epidemic is as shown in the table below. The third row of the table shows the ratio of the number of cases in each week to the previous week:

Week number (t)	0	1	2	3	4	5
Number of new flu cases, N_t	56	100	182	329	586	1055
$\frac{N_t}{N_{t-1}}$	–	1.79	1.82			

a) Complete the rest of the third row of the table (rounding to three significant figures).

b) There are claims that the number of cases is growing 'exponentially'. Explain why this appears to be true.

c) In week 6, the number of new cases is reported as being 1594. Do doctors have any reason to feel optimistic?

ANSWERS

Q1

a) The price has risen by 9p. The increase is $9 \div 60$, which is 0.15 or 15%. The temptation for most adults would be to reach for a calculator, but teenagers are encouraged (or, in non-calculator papers, forced!) to look for simple shortcuts. Typically they will simplify the fraction to $\frac{3}{20}$ and then see that this is $\frac{15}{100}$ or 15%.

b) The saving is £10, as the original price was £50. Price reduction questions are a notorious stumbling block – the majority of teenagers see the numbers £40 and 20% and automatically work out 20% of £40 (which is £8). But calculating '20% off' means multiplying a price by 0.8 – for example, 20% off £100 is $0.8 \times 100 =$ £80. So the actual calculation required here is: original price $\times$ 0.8 = £40, hence the original price = £40 $\div$ 0.8 = £50. For a discussion of problems like this, see the section on Ratios and Percentages that starts on page 47.

Q2

2.4×10^4 . Although 24×10^3 is a correct number, the rule is that standard form numbers must be between 1 and 10. We discuss standard form on page 210

Q3

a) 3432. Most teenagers will spot this is the same calculation but with the numbers scaled up or down by tens and hundreds. Their most likely error will be putting the decimal point in the wrong place. The safest way to check is by doing a rough approximation. The calculation is very roughly 800×4, which is 3200, so we are looking for an answer around 3000.

b) 34.32 This question tests out how comfortable your teenager is with doubling and halving numbers, and realising that if you double one number and halve the other, the product is the same. The final twist is that there is a shift in the decimal point. Again, a rough estimate helps to see that the answer will be close to $4 \times 9 = 36$.

Q4

L = 40. By far the most common error here is to reason that since 15 was increased by 10 to 25, L must also be increased by 10 to give the answer 34. No calculator is needed here if you spot that the ratio of 24:15 is the same as 8:5, and that this is the same as 40:25.

Q5

15 more white balls are needed. This question requires confidence in understanding what ratios mean. A teenager who is shaky on ratios might well take the ratio 1:4 as literally meaning that there must be one white and four blacks, and that a 4:1 ratio must therefore mean there are four white balls. What 4:1 white:black does mean, of course, is that there are four times as many whites as blacks. Since there are 4 blacks, there need to be 16 whites. Many teenagers, relieved at working out 16, will have forgotten that the question asked how many more whites are needed and thus write 16 instead of 15.

Q6

AC (the hypotenuse) is 8 metres. Solving this means you have to know how to use the standard equations from trigonometry (see page 166). In this case, AB is opposite the angle that you know, so this is a calculation involving sine.

$$\sin(30^\circ) = AB \div AC$$
$$0.5 = 4 \div AC$$
$$AC = 8$$

Pupils find it harder to work out the hypotenuse than the other sides because it appears in the denominator (i.e. the bottom of) the fraction. Just as with the percentage reduction question 1(b), having found that $\sin(30) = 0.5$ they will often then multiply 4×0.5 to get an answer of 2 metres. With trigonometry questions, it's useful to estimate what the answer might be before doing the calculation. Since we're working out the longest side of the triangle, it must be longer than 4 metres.

Q7

$x = 5, y = 4$

There are at least two ways of solving this. Older teenagers will recognise that these are simultaneous equations (see page 107) and many will enter auto pilot by solving them using methods they have been taught (and in some cases, getting tangled in the process).

One approach is to divide the second equation by 10, so it becomes $3x - 3y = 3$. Now the two equations can be added together to eliminate y:

$$7x + 3y - 3y = 35$$
$$7x = 35$$
$$x = 5$$

Substitute $x = 5$ into the first equation:

$$20 + 3y = 32$$
$$3y = 12$$
$$y = 4$$

On the other hand you – like some younger teenagers – might well take a more straightforward approach by spotting in the second equation that if $30x - 30y = 30$ then $x - y = 1$. A quick bit of trial and error with the first equation reveals that $x = 5$ and $y = 4$ works.

This is a reminder that common sense, and 'homespun' methods, can be just as effective as learned methods – and it's great if you are able to use both.

Q8

$a = 10\text{cm}, b = 2.7\text{cm}$

The area calculation is $5 \times a = 50$, so $a = 10$, and $10 = 7.3 + b$ so $b = 2.7$. The maths here is quite easy, what is unusual, and for many teenagers challenging, is that this question is 'back to front'. Usually the challenge is to work out the area of a rectangle by multiplying the two sides, but here they are being asked to work out one of the sides using the area. The question didn't need to ask for the value of a, but working out a is an essential stepping stone to working out b.

The other confusion that might arise is between area and perimeter (many teenagers get these two muddled). If the perimeter of the rectangle were 50cm, then a would be 20cm and b would be 12.7cm.

Q9

$y = 13$. This could be solved by trial and error: $y = 10$ gives an answer that is too small, $y = 20$ gives one that is too big. But solving it by rearranging the equation is much quicker:

$$
\begin{aligned}
2(y+1) \div 7 &= 4 \\
2(y+1) &= 28 \\
(y+1) &= 14 \\
y &= 13
\end{aligned}
$$

Thinking of equations as a balance can help here, see page 104.

Q10

$a = 40^{0.}$ Is this a geometry question or an algebra question? In fact, it is both. Mathematical topics often overlap, and questions like this are testing teenagers to check that they can work across the boundaries. Many teenagers find this difficult, especially if they are taught topics in separate modules without ever seeing connections made between one topic and the next.

Here, they need to spot that the angles XPR + SQY = 180º.

In other words, $3a + (a + 20) = 180$.

Now it's a simple algebra problem:

$$4a + 20 = 180$$
$$4a = 160$$
$$a = 40$$

Q11

a) Four of the 20 socks are white, hence the chance of drawing one at random is $\frac{4}{20}$, which is more simply expressed as $\frac{1}{5}$. Probability problems are the most common 'real world' situations where a teenager will be expected to know how to simplify fractions.

b) Having picked out one white sock, the total number of white socks remaining is 3 and the total number of socks left in the drawer is 19, so the probability that the second sock is white is $\frac{3}{19}$.

A typical error is to assume that the chance of getting a particular colour will never change (just as the chance of getting heads on a coin is always the same). One way to see why the probability of getting a particular sock colour changes as you remove socks from the drawer is to imagine there are just two socks, one black and one white. The chance of the first sock that is removed being white is ½. But since that is the only white, the chance of the second also being white cannot be ½. In fact it is zero!

Q12

The minimum number of marbles, x, is 7. This problem is an algebraic equation wrapped up in a different context. The reason why examiners sometimes do this is to force teenagers to think rather than just blindly following a procedure. The biggest challenge here is in turning the wordy problem into a mathematical expression. Since box A contains more marbles than box B, what we have is an inequality:

$$5x + 2 > 3x + 14$$

Inequalities can be rearranged and simplified in exactly the same way as any equation, by doing the same to both sides. So:

$$5x + 2 > 3x + 14$$
$$5x > 3x + 12$$
$$2x > 12$$
$$x > 6$$

In other words there are more than six marbles, and so the minimum number of marbles that fits the conditions of the question is seven. This can easily be checked. With $x = 6$ there are 32 marbles in box A and 32 in box B, but with $x = 7$ there are 37 in box A and 35 in box B.

Q13

a) $12x^2 - 2$. As we point out on page 86, expand is a strange word since the end result doesn't exactly seem expanded, if anything it has got smaller. It's just one of those words teenagers have to get used to.

Questions like this require confidence in multiplying negatives, and in knowing how to treat brackets – much less daunting if you see the connection with the grid method (page 87). Here's how to do it in this example:

$$(3x + 2)(4x - 1) - 5x$$
$$= 12x^2 - 3x + 8x - 2 - 5x$$
$$= 12x^2 - \cancel{3x} + \cancel{8x} - 2 - \cancel{5x}$$
$$= 12x^2 - 2$$

Incidentally, this is a quadratic (see page 98), though no mention is made of this in the question.

b) $3x^2 + 7x + 2 = (3x + 1)(x + 2)$. Most parents struggle to remember what 'factorise' means (while their teenagers, drilled in how to do it, still struggle to understand what it is for). Think of factorising as the opposite of expanding, in other words find the expressions in brackets that when multiplied together give this result. The grid method approach (page 87) is one way to factorise, though teenagers learn other methods too.

c) $3x^2 + 7x + 2 = 0$ when $x = -2$ or when $x = -\frac{1}{3}$. Using the answer from (b), the question is asking to find values of x where $(3x + 1)(x + 2) = 0$. If $(3x + 1)$ multiplied by $(x + 2)$ is zero, it means that either $3x + 1 = 0$ (so $x = -\frac{1}{3}$) or $x + 2 = 0$ (so $x = -2$).

This question could have used much blunter language: 'Solve the quadratic equation $3x^2 + 7x + 2 = 0$'. It means the same thing.

Q14

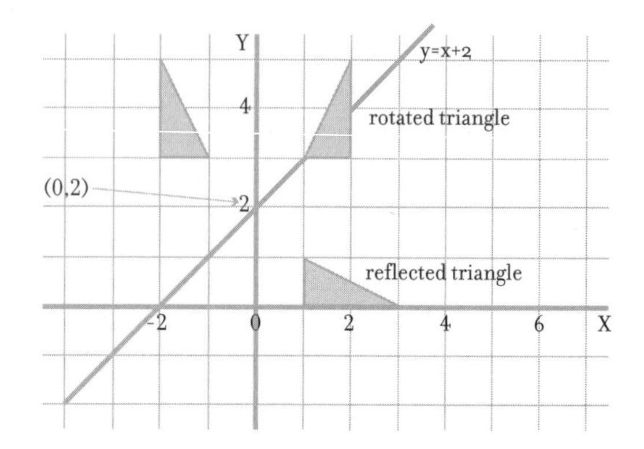

Treated separately the three parts of this question are very standard questions that a Year 9 pupil might be faced with. What is more unusual is combining the plotting of a graph with doing a transformation of a shape.

The most common error in these questions is the position of the reflected and rotated triangle. It's always easier to picture where the reflected position will be if you turn the page so that the line of reflection ($y = x + 2$) is vertical – then you can plot each of the triangle's corners in turn, making sure the original and the reflection are the same distance from the line of reflection.

One way to rotate the triangle is to draw lines from each corner to the centre of rotation, (0,2). Then rotate those lines by 90° to show the new positions of the corners. Alternatively use some tracing paper, as many teenagers do.

Notice how the final rotated triangle is just a reflection of the original triangle in the y-axis (which is the line $x = 0$). This is the answer to part (d) of the question.

There's more about reflection and rotation on pages 178–182.

Q15

The smallest counter-example is 7, because 7 is a prime number but $2 \times 7 + 1$ is 15 which is not prime. (Other examples include 13 and 17.) There's no hard maths here, but many teenagers are daunted by the language of maths questions, such as 'if n represents a prime number . . .'

The question could have been written like this:

Write the first five prime numbers in the first row of this table. Then fill in the rest of the table.

First Row: *Prime numbers*					
Second Row = *Double First Row*					
Third Row = *Second Row + 1*					

Are any of the numbers in Row 3 NOT a prime number?

In this form, any teenager who knows what prime numbers are could tackle the problem easily. (The first row is 2, 3, 5, 7, 11, the second is 4, 6, 10, 14, 22 and the third 5, 7, 11, 15, 23.) Your teenager may well need help and encouragement in turning problems written in maths language into more everyday language. Incidentally, this question is a simple example of proof, a feature of secondary maths that is being given increasing attention after (some people think) years of neglect.

Q16

Maths exam questions have a habit of presenting familiar problems with an unusual twist thrown in. This is a 'circle theorem' question.

a) The reason why OPR forms an isosceles triangle is that the lines OP and OR are both the radius of the circle, so they are the same length. A triangle with two sides that are the same length is isosceles.

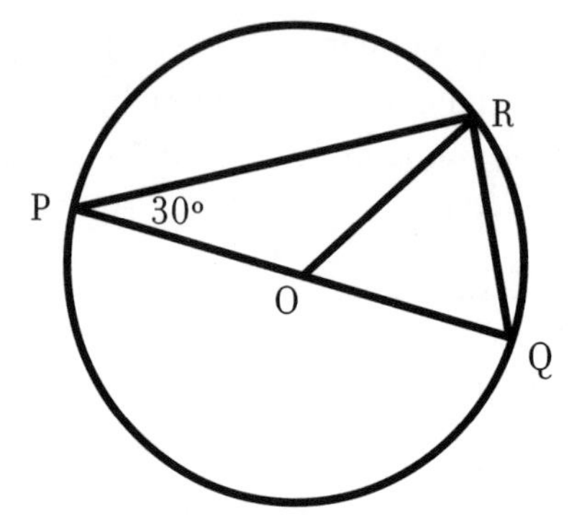

b) The reason why triangle ORQ is equilateral is that its three angles are all 60°. This question tests the knowledge of a circle theorem that teenagers are expected to memorise. The theorem says: 'The angles formed by drawing lines from the ends of the diameter of a circle to its circumference always form a right angle.' In other words the theorem says that angle PRQ is 90°. From this the other angles can all be deduced:

Since triangle OPR is isosceles, the angle OPR = angle ORP = 30°
Hence angle ORQ = 90° – 30° = 60°
Length OR = Length OQ so angle ORQ = angle OQR = 60°
The three angles in triangle OQR add to 180° so angle QOR is also 60°

As a parent you probably don't know the right-angle theorem (sometimes called Thales' theorem). However, it turns out you don't have to know it at all – in geometry it's always possible to work out answers by going back to the very basic theorems of triangles and circles that we introduced in the first geometry chapter on page 139 – it just takes longer. In the example above we want to show that angle ORQ (which we've called x below) is 60°. We've already shown in part (a) that triangle OPR has two 30° angles. Since triangle OQR is also isosceles (OR and OQ are the same length) angles x and y must be the same as each other. The internal angles of triangle PRQ add to 180° so:

$$
\begin{aligned}
30 + 30 + x + x &= 180 \\
60 + 2x &= 180 \\
2x &= 120 \\
x &= 60
\end{aligned}
$$

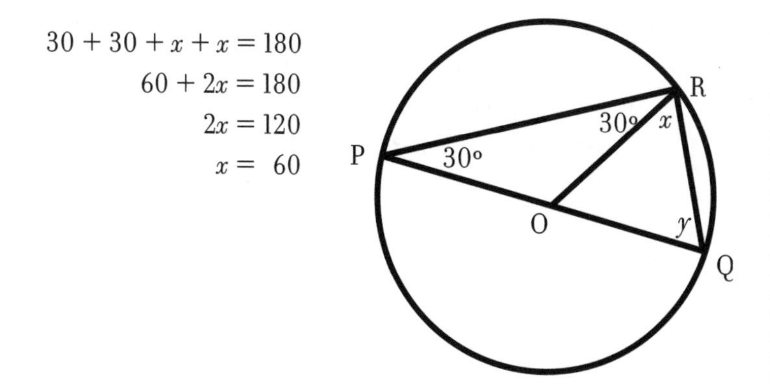

It's a bit of a slog, but the beauty (if you want to call it that) of geometry is that you can deduce everything by going back to first principles.

Q17
Graph interpretation questions can really tax some teenagers, as we discussed in the graphs chapter (page 114).

a) The car was filled with petrol at the three-hour mark. The line looks vertical but in reality it has a very slight slope as filling up takes a minute or so.

b) The car was stationary for about one hour. This question is a typical one for testing the ability to interpret what graphs mean. When the car is parked the graph is horizontal (no petrol is consumed as time elapses). Both horizontal segments of the graph are about half an hour.

c) The kink in the graph after $4\frac{1}{2}$ hours indicates that the car begins to consume more petrol per unit of time. There are several ways that a car's fuel consumption rate can increase, but the most likely is that the driver began to drive much faster after $4\frac{1}{2}$ hours, so that it was consuming more petrol per minute. Since the graph is a straight line either side of the kink, it seems that the car changed from one steady speed to a faster one, for example, by moving off an A road onto a motorway.

Q18

Only (b) could possibly be true. This question is exploring the ability to think about the properties of negative numbers in an abstract way, since we are never told what the two numbers are. Many teenagers would freeze at a question like this. The way to help them tackle it is to remind them that one of the top tips in tackling maths problems is to try it out with a couple of simple numbers and see what happens.

What if both numbers are negative, say $^{-}3$ and $^{-}4$? $^{-}3 \times {}^{-}4 = 12$. Indeed when you multiply any two negative numbers, their product is always a positive number, so (a) cannot be true.

For the two numbers to have a negative sum and a negative product, one must be negative and the other positive, $^{-}10$ and 5, for example.

It's possible for (b) to be true, though only two solutions would work: $^{-}4$ and 1, or $^{-}3$ and 2.

Is it possible to double one of the numbers and get the other? No, because twice a negative is still negative and twice a positive is still positive ($5 \times 2 = 10$, not $^{-}10$). So (c) cannot be true.

Could (d) be true? No, because if the squares of the numbers are the same, then the two numbers must be the same magnitude, for example $^{-}4$ and 4, or $^{-}10$ and 10, so they add to zero.

Q19

(a) 50%.

Two of the four numbers on the dice are multiples of five (5 and 10) so the chance of one dice landing on a multiple of 5 is $\frac{2}{4}$ or 50%.

(b) The chance of the product of the two dice being a multiple of 5 is 75%.

Most pupils would want to methodically write out all of the possible combinations in a table:

×	2	5	7	10
2	4	10	14	20
5	10	25	35	50
7	14	35	49	70
10	20	50	70	100

Of the sixteen possible outcomes, twelve are multiples of 5, so the answer is $\frac{12}{16}$, which is $\frac{3}{4}$ or 75%.

However, there is a quicker method for those comfortable with factors. If you roll the two dice, and either of them ends up on a multiple of 5, then the product will also be a multiple of 5. This is obvious if you think of an example: 5 times anything is a multiple of 5, as is 10 times anything and 15 times anything. So the only time when the product won't be a multiple of 5 is when neither dice comes up as a multiple of 5, which is 50% × 50% = 25%, or $\frac{1}{2} \times \frac{1}{2} = \frac{1}{4}$. In other words, for $\frac{3}{4}$ (or 75%) of the time, the product will be a multiple of 5.

In fact the maths here is exactly the same as working out the chance of getting at least one head when two coins are flipped. Done as coins, it's a regulation, straightforward probability question. But present it as multiplying numbers on dice, and most people regard this as fiendishly difficult.

Q20

This is an example of the more challenging functional maths questions that your teenager might encounter these days – using mathematical knowledge to interpret a real-life situation.

(a) Here is the completed table:

Week number (t)	0	1	2	3	4	5
Number of new flu cases, N_t	56	100	182	329	586	1055
N_t / N_{t-1}	-	1.79	1.82	1.81	1.78	1.80

Teenagers often get confused between significant figures and decimal places. The numbers in the third row are accurate to two decimal places but three significant figures (see page 208 for a reminder).

(b) Growth is exponential when each number in the sequence is a fixed multiple of the previous number. In this case, the ratio of flu cases to the previous week is always 1.80 (plus or minus a tiny amount) so the number of cases is indeed growing exponentially. See more about exponential growth on page 131.

(c) If the number of cases in week 6 is 1594, then the rate of growth from week 5 is $1594 \div 1055 = 1.51$. This is a sharp drop from the 1.80 in the previous week and a sign that the epidemic is no longer growing exponentially. So yes, even though the number of cases has increased significantly during the week, the doctors have reason to be cautiously optimistic that the corner is being turned.

ANSWERS TO TEST YOURSELF QUESTIONS

PAGE 43

a) $19 + 63 \approx 20 + 60 = 80$
b) $38 \times 12 \approx 40 \times 10 = 400$
c) $1.84 \times 97 \approx 2 \times 100 = 200$
d) $436 \times 68 \approx 400 \times 70 = 28{,}000 \approx 30{,}000$

PAGE 53

1) $\frac{3}{4} = \frac{9}{12}$. $0.75 = \frac{?}{12}$ so $? = 0.75 \times 12$. (Alternatively, notice that the top and bottom of the fraction have both been multiplied by 3.)
2) $6\frac{1}{4}$ (or 6.25). $\frac{5}{?} = 0.8$, so $? = \frac{5}{0.8}$.
3) 50. Written as fractions the question becomes $\frac{6}{20} = \frac{15}{?}$.
4) You can make tea for just over 100 people (or $106\frac{2}{3}$ if you want a precise answer). The shorthand way of working this out is to write it as: $\frac{3}{4} = 0.75 = 80 \div P$ where P is the number of people you can cater for. So $P = 80 \div 0.75 = 106.67$. However, that is certainly not the only way to work out the answer, and when problems are presented in familiar real-world situations, it's common for adults to use

more intuitive methods (maybe that is what you did). And that's fine!

PAGE 56–7

1) 10% of 350 is 35, so 40% is 4 × 35 = 140.
2) Inadvertently (we hope) the shop ended up diddling the customer. (Let's say the wine cost £100. A 20% discount would have made it £80, whereas a 10% discount reduces it to £90, and a second 10% discount from £90 brings it down by another £9 to £81.)
3) Both end up on the same wage. Increasing a wage by 20% is the same as multiplying it by 1.2. Reducing by 10% is the same as multiplying by 0.9. Kate's wage was therefore multiplied by 0.9 × 1.2, while Jasmine's was multiplied by 1.2 × 0.9. Since the order of multiplication makes no difference, the end result is the same.

PAGE 78

a) Both. This is the formula for working out any temperature in Fahrenheit if you know the temperature in Celsius. It is also an equation because all formulae are equations!
b) Equation. This tells us the connection between Sally and Bobby's ages. It is not a formula, however, because it's not a general rule for anyone called Sally and Bobby, it's a connection between these two people that only applies at their current ages.
c) Neither. You probably recognise that this is the expression for the area of a circle. To be an equation (and formula) it would have to be written with the equals sign as: Area = πr^2.

d) Equation, though it is a meaningless one, since there is no number for which it is true. The equation is effectively saying '$3 = 4$', which is nonsense.

PAGE 81

a) The fifth term is $(2 \times 5) + 5 = 15$. The 105th term is 215. In effect, you are being asked, 'What number is in the fifth pigeonhole, and what number is in the 105th?' The rule is to multiply the number of the pigeonhole by 2 and add 5.
b) The 10th term is 69. There's nothing wrong with plugging $n=10$ into each part of the expression, $4 \times 10 + 3 \times (10 - 1) + 2$. However, it's quicker if you simplify the expression to $4n + 3n - 3 + 2 = 7n - 1$. You can now quickly get to the answer $(7 \times 10) - 1 = 69$. There is more about simplifying algebra on page 93.
c) Here you are being asked to find what the rule is for each pigeonhole, which is much more of a challenge than just following the rule that you have been given. Looking down the list you may spot the pattern. For each pigeonhole you multiply the number of the pigeonhole by 7. Then you add to that 'the number that is one larger than the pigeonhole number, *multiplied by 3*', And to that you add 5. Written out mathematically, the rule is: $7n + 3(n + 1) + 5$. That will do as an answer, though there is a simpler way of saying the same thing, which is $10n + 8$.

PAGE 82

$6P = S$. The temptation is to write the algebra exactly as the sentence reads, 'Professors $= 6 \times$ Students' but a university with 600 professors and only 100 students would be under severe scrutiny from a government looking to cut back on over-manning.

PAGE 92

a) $7a + 7b$

b) $x^2 + 3x + 4x + 12$; this can be further simplified to $x^2 + 7x + 12$

c) $8a + 2a^2$

d) This is the same as $(a + b)(a + b)$, and the answer is $a^2 + 2ab + b^2$

e) $2ax + bx + 3x - 2a - b - 3$. There are six terms in this expression. One way to check you have the right number of terms (before you do any simplifying) is to multiply together the number of terms there are in each bracket. Here we had two terms in the first bracket $(x - 1)$ and three in the second $(2a + b + 3)$ which meant there *must* be $2 \times 3 = 6$ terms in the answer. This example included a multiplication by negative numbers – there's more about this on page 203.

PAGE 103

1a) $19^2 = 361$ Simplify it by adding and subtracting 1, to give 20×18 (= 360) and then add 1 to get 361.

b) $101^2 = 10201$ Again, simplify it by adding and subtracting 1, to give $100 \times 102 = 10200$ and then add 1 to get 10201.

2a) $3(a + b)$

b) $2(1 + 4q^2)$

c) $(x + 4)(x + 1)$ – solved in just the same way as the example on page 99.

d) $(x + 6)(x - 1)$ – this can also be solved using the grid method, but it has the extra catch of involving negative numbers. We're looking for a solution of the form $(x + a)(x + b)$.

The 'clues' are that we know $ab = {}^{-}6$ and that $a+b = 5$. If $ab = {}^{-}6$ then either a OR b is a negative number, but not both (as two negatives multiplied together make a positive). Some trial and error leads to the discovery that a must be 6 and b is $^{-}1$. To learn more about multiplication of negative numbers and the common pitfalls, see page 203.

PAGE 110

a) The hat costs £60 (and the coat £240). Written as equations, if you call the price of a hat h and the price of a coat c then: $h + c = 300$ and $c = 4h$. So $h + 4h = 300$, so $5h = 300$, so $h = 60$.

b) $x = 7$ (and $y = 1$). This question is simple enough that you don't really need to know about simultaneous equations in order to solve it. However, we've sneakily thrown in negative symbols here to flag up that manipulating negative numbers throws many teenagers into turmoil. The obvious thing to do with the equations is to eliminate the 'y', but what is $(3x - y)$ minus $(x - y)$, do you add or subtract? In fact $3x - y - (x - y) = 2x - y + y = 2x$, easily tested by calling x and y the numbers 2 and 1 (say), but a strong understanding of arithmetic with negative numbers is essential in algebra. See pages 199–205.

PAGE 111

a) For 100 minutes in a month, MegaCall costs £20 + 0.1 × 100 = £30. NanoPhone costs £10 + 0.15 × 100 = £25. So MegaCall costs more.

b) The monthly cost in pounds of MegaCall, $M = 20 + 0.1t$ (where t = time spent on the phone in minutes); the monthly cost of NanoPhone, $N = 10 + 0.15t$.

c) The two companies charge the same when $M = N$, so:

$$20 + 0.1t = 10 + 0.15t$$

Subtract $0.1t$ from each side: $20 = 10 + 0.05t$

Subtract 10 from each side: $10 = 0.05t$

Multiply both sides by 20: $200 = t$

So the two companies charge the same amount when the monthly usage is 200 minutes.

PAGE 123

a) The graph will look roughly like this:

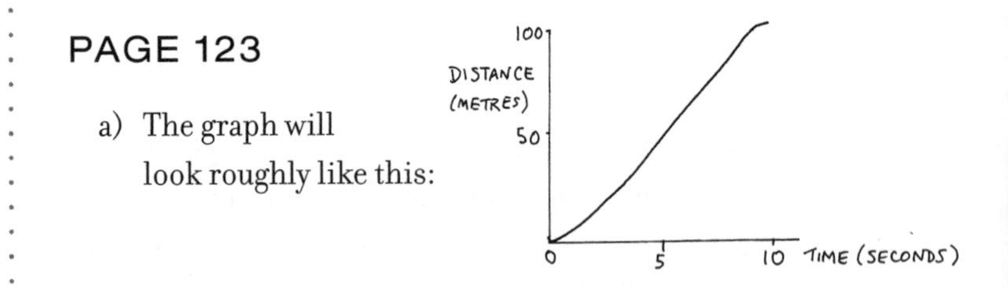

It is roughly a straight line (though at the start it will be a bit flatter as he builds up speed).

b) The slope would be just under 100 metres *per 10 seconds*, or 10 metres per second. When you plot a graph of distance travelled against time, the slope of the graph always represents the speed at which the object was travelling.

PAGE 154

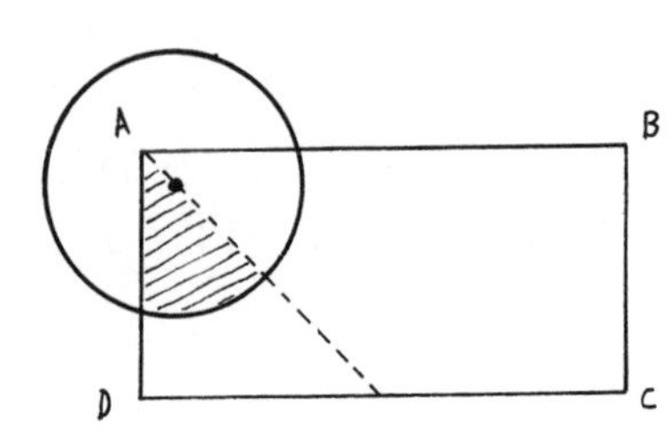

This is a typical locus problem that teenagers might face, as it involves two separate conditions.

The circle indicates all points that are within 2 metres of the tree.

The dotted line at 45° from A is all the points that are the same distance from AD and AB (the bisector of those two fences). The points that are *nearer* to AD are all the points in the triangle that is below and to the left of the dotted line.

So the flower bed is the shaded area. It is the locus of points that meet BOTH of the criteria (within 2 metres of the tree AND nearer to fence AD than to AB).

Finding intersecting regions like this is reminiscent of creating Venn diagrams, something that your teenager first encountered in primary school and which is a technique often used when sorting data.

PAGE 157

Two sides AE and BC are the same length even though they have been drawn inaccurately (they both have the single line cutting them to indicate that) and they are also parallel (the chevrons). CD and DE are also the same length as each other. Because of the parallel lines, the angles at A and B are Cangles. These therefore add to 180°.

PAGE 164

The dodecagon can be divided into ten triangles, for example like this:

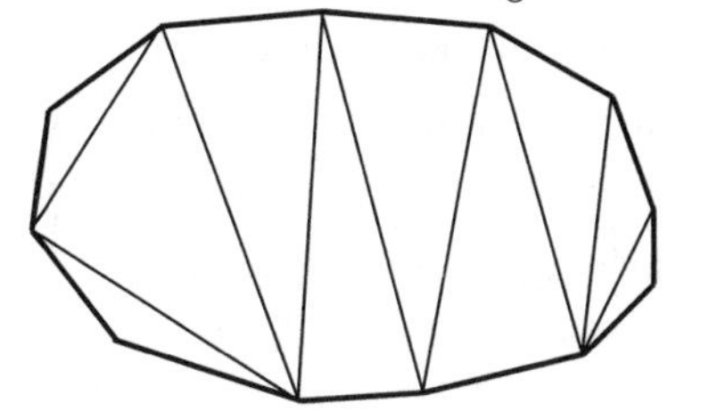

So the internal angles add to $180 \times 10 = 1800^{\circ}$.

And the internal angle of a regular dodecagon is therefore $1800 \div 12 = 150^{\circ}$.

PAGE 171

$AE \div AC = AD \div AB$. Putting in the numbers, $8 \div 5 = AD \div 7$. The easiest way to find AD is to multiply both sides of the equation by 7, so $AD = 7 \times 8 \div 5 = 56 \div 5 = 11.2$.

PAGE 175

a) BC is Opposite the angle and we want to know the Hypotenuse, so we have: Sine = O ÷ H or $\sin(40) = 10 \div AB$. From a calculator, $\sin(40) = 0.643$. Rearranging, $AB = 10 \div \sin(40) = 15.6$cm. Commonly, teenagers would make the mistake in this problem of working out AB multiplying $\sin(40) \times 10$. This is yet another example of why it's essential to know that if $A = B \div C$ then $C = B \div A$ (see page 49).

b) We know the Hypotenuse (200cm) and the side that is Adjacent to the corner (25cm), so this is a cosine calculation, Cosine = A ÷ H = $25 \div 200 = 0.125$. Use the $\cos^{-1}$ or INV cos button on your calculator to find the angle whose cosine is 0.125, which is 82.8°.

PAGE 187

The nine dotted lines indicate all the lines of symmetry of a cube. If you need convincing that there aren't any more, get a large cube of cheese and cut along the lines using wire.

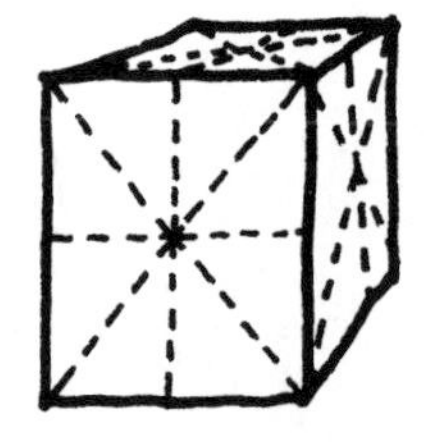

PAGE 190

If visualising rotations isn't your thing, it isn't immediately obvious. In fact, the centre of rotation is C, the position with co-ordinates (5,2). You can confirm that the distances from C to the tip and other corners of both triangles are the same. You can also use the tracing paper and drawing pin method or the mathematical construction method (page 181) to confirm it.

PAGE 204

a) $^{-}19$. Start at $^{-}7$ on the number line, and then move 12 in the negative direction.
b) $^{+}16$. Subtracting a negative is the same as adding a positive.
c) $^{+}21$.
d) $2a - 7$. Multiplying out the brackets we get $3 - (2 \times 5) - (-2a) = -7 + 2a$. It's usually easier to read expressions with the positive part first, so the best answer is $2a - 7$.

PAGE 211

a) 2.49×10^4
b) 4.6×10^9 (do the calculation 2×2.3 first, then add the powers of ten.)
c) 3×10^2

PAGE 215

a) 0.342 is larger. The reason we pose this question is that teenagers often think that the 'longest' decimal must also be the largest. (They can also make the opposite mistake, and think that because a number contains digits in the 'ten thousandths' column, it must be smaller than a number with no digits in that column, so they might incorrectly think that 0.3282 is smaller than 0.328.)

b) 0.042 is 4.2% while 1.07 is 107%. There's no rule to say percentages cannot be larger than 100 – for example, prices have been known to grow by hundreds or thousands of per cent in countries suffering severe inflation. However, there are many situations where 107% would be meaningless, for example, when describing the results in an examination or the level of commitment given by a sportsman to his team, 'The boys gave it 110% today!'

PAGE 224

a) 3.09×10^{-2} Count how many places the decimal point has to be moved $0.\underbrace{03}_{1\,2}0937\ldots$

b) 8.12×10^{-4}

PAGE 226

a) $\frac{3}{5}$. Multiply the top and bottom numbers in the fractions to get $(3 \times 4) \div (4 \times 5) = \frac{12}{20}$. You can simplify the calculation by spotting that the 4s on the top and bottom can be cancelled out.

b) $\frac{14}{27}$ (fourteen twenty-sevenths). The word 'of' is shorthand for 'multiplied by', so written as numbers this is $(\frac{2}{3}) \times (\frac{7}{9}) = \frac{14}{27}$.

c) Don't make the common teenager error of thinking 'divided by one third' is the same as 'divided by three'. Think of this as 'How many thirds are there in one?' There are three, so there must be $3 \times 6 = 18$ thirds in 6.

d) $\frac{4}{3}$. For calculations like this, it pays to have memorised the rule, *turn upside down and multiply*. $\frac{5}{6} \div \frac{5}{8} = \frac{5}{6} \times \frac{8}{5} = \frac{40}{30}$, or $\frac{4}{3}$.

PAGE 236

The room's dimensions are recorded as 430cm × 310cm, but these numbers are rounded to the nearest 10cm. The dimensions of the room could therefore have been as large as 434.9cm × 314.9cm (or 4.35m × 3.15m) = 13.7 square metres. That's 0.4 square metres more than the original answer of 13.3 square metres. In other words, there could end up being a gap equivalent to a piece of carpet that is 0.4 square metres, or 40cm by 1 metre – surprisingly large. In situations such as carpeting or buying paint it makes sense to round measurements up so that you err on the large side – otherwise you'll find yourself on an unwanted return visit to B&Q.

PAGE 237

c) Peter can't catch Rita.

Most teenagers' (and adults') intuition says that if Rita cycles at 10km/hour for the whole 20km journey, then Peter must have the same average speed, so if he travels up the hill at 5km/hour, he has to travel down at 15km/hour (because $(5 + 15) \div 2 = 10$). But a little thought reveals that this cannot be true. Rita's journey took two hours. If Peter cycles at 5km/hour then it takes him two hours just to do the uphill part of the journey. That means he has to travel the next 10km in zero time if he is to catch Rita! Confused? The problem here is that speed is a compound measure, which means you can't apply the rules that apply to taking the average of something like distance or weight.

PAGE 239

1) $C = (F - 32) \div 1.8$.
2) Centigrade and Fahrenheit are the same at -40°. In 2011 Prince Harry went on an Arctic expedition and was caught in -40° conditions. For once it didn't matter that the headline writers didn't name the units of measurement.

PAGE 244

a) The difference in the circumference is 2π, or about 6.28 metres. (The circumference of the larger circle is $\pi \times 4$ while the smaller is $\pi \times 2$).
b) Kelly Holmes ran about 2π (that's 6.28) metres further. In fact this is exactly the same answer as for (a)! The width of an athletics lane (about one metre) is the extra radius of the circular track, so the extra diameter of the lane that Kelly Holmes ran in was two metres. What is surprising is that the answer is the same whatever the radius of the inside lane curve is. If we call it R, then the inside lane's circumference is $2\pi R$. The second lane's circumference is therefore $2\pi(R + 1)$. Work out the difference: $2\pi(R + 1) - 2\pi R = 2\pi R + 2\pi - 2\pi R = 2\pi$!

PAGE 249

a) There are 100 centilitres in a litre, just as there are 100cm in a metre.

b) Since there are 100cm in a metre, there are 100 × 100 × 100 = 10^6 (one million) cubic centimetres in a cubic metre.

c) The cubes each have a volume of 20 × 20 × 20 cubic cm, which is 8000cc, or 8 litres. So at most 100 of these cubes could be fitted into the vault.

d) The area of the cylinder is pi × its radius squared, and to get the volume, multiply by its height '*a*' to get: pi.z.z.a ☺

PAGE 277

a) **Lucky red.** The chance of winning is $\frac{1}{4}$. The chance of tossing a head is $\frac{1}{2}$ and the chance of drawing a red card is $\frac{1}{2}$. The chance of doing both can be found simply by multiplying the two probabilities together, but only because the two probabilities are independent.

b) **Weather forecast.** The correct answer is 75% – though strictly speaking this is only true if what happens on Saturday has no bearing on what happens on Sunday; in other words, Saturday and Sunday weather forecasts need to be independent of each other. (Weather forecasting is so hit and miss that even quoting a specific probability of rain is a bit dodgy.)

Let's explain where the answer of 75% comes from. Since we've been told the probability of rain on each day is 50%, let's think of this in terms of flipping a coin, we will call 'heads' to represent a day when it rains and 'tails' if it's a dry day.

The four possible outcomes are:

Saturday	*Sunday*
Rain (heads)	Rain (heads)
Rain (heads)	Dry (tails)
Dry (tails)	Rain (heads)
Dry (tails)	Dry (tails)

There are four possible outcomes, and each of these outcomes is equally likely. Only one of the possible outcomes, the last one, is completely dry. So the chance of a completely dry weekend is one in four, or 25%, and the chance that there's rain on at least one day is three out of four, or 75%.

But very few teenagers or adults come up with this answer. In the case of the weather forecaster, he cheerfully announced: 'It's a 50% chance of raining on Saturday and 50% on Sunday so there's a 100% chance that you'll be getting some rain this weekend.' He simply added the probabilities together. You can see why this is nonsense if you imagine it was a Bank Holiday weekend with the forecast of a 50% chance of rain on Monday as well. According to this weather forecaster's maths that would mean a 150% chance of rain over the holiday, a meaningless number.

Most teenagers, on the other hand, go in the other direction, saying that the chance of rain over the weekend is 50%. Why? Here's how one girl explained it: 'There's a $\frac{50}{100}$ chance of rain on Saturday and $\frac{50}{100}$ chance on Sunday, so that's $\frac{100}{200}$ for both days and that's still a half, or 50%.'

You can see where her reasoning came from. Suppose you take an exam and score $\frac{50}{100}$ and then you take a second exam and get the same score of $\frac{50}{100}$. Over the two exams you score $\frac{100}{200}$, or 50%. The girl used this sort of reasoning for the weather forecast. At first glance it sounds convincing, but probability doesn't work in the same way as adding up exam marks. A simple experiment where you flip a coin to represent rain over the two days and collect the results will demonstrate that it is wrong. But *proving* why 50% is the wrong answer to a confused teenager is challenging, and this is an example of why probability can (for many teenagers) become a particularly difficult topic to grasp.

PAGE 278

1) You should bet on horse number 7 as that's the most likely to win. The total score of two dice can be anything from 2 to 12, so horse 1 will never leave the starting gate. To see why the other totals are not equally likely, you can list all of the possible outcomes of the two dice, which we'll label as the red dice and the black dice. All the combinations are listed here, with their totals in brackets:

Red, Black	Red, Black	Red, Black	Red, Black	Red, Black	Red, Black
1 + 1 (2)	2 + 1 (3)	3 + 1 (4)	4 + 1 (5)	5 + 1 (6)	6 + 1 (7)
1 + 2 (3)	2 + 2 (4)	3 + 2 (5)	4 + 2 (6)	5 + 2 (7)	6 + 2 (8)
1 + 3 (4)	2 + 3 (5)	3 + 3 (6)	4 + 3 (7)	5 + 3 (8)	6 + 3 (9)
1 + 4 (5)	2 + 4 (6)	3 + 4 (7)	4 + 4 (8)	5 + 4 (9)	6 + 4 (10)
1 + 5 (6)	2 + 5 (7)	3 + 5 (8)	4 + 5 (9)	5 + 5 (10)	6 + 5 (11)
1 + 6 (7)	2 + 6 (8)	3 + 6 (9)	4 + 6 (10)	5 + 6 (11)	6 + 6 (12)

There are thirty-six combinations in total, but only one combination adds to 2 (the one in the top left of the grid, 1 +

1), and only one adds to 12 (6 + 6). On the other hand there are six ways of combining two dice to score 7, making this the most likely score. In fact the chance of scoring 7 is $\frac{6}{36}$, or 1 in 6.

The dotted line in the chart shows the theoretical pattern that the dice (or horses) will make at the end of the game, while the positions of horses themselves are what you might typically get in practice because of the random nature of the outcomes.

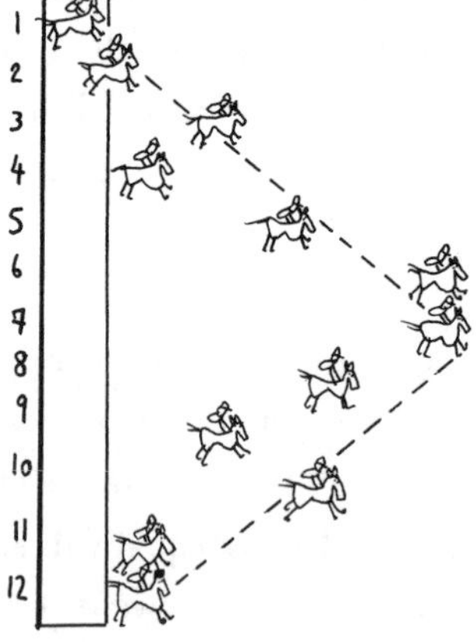

2) The strategy that would give you the highest score if you played this game hundreds of times is to 'stick' as soon as your score reaches 15. A GCSE pupil wouldn't be expected to be able to do the maths to prove this, but that doesn't matter because they will still engage with the game, and come up with their own theories on the best strategy. Teenagers' tactics vary widely, from those who are very risk averse and bank after the first throw to those who keep going even when they get to scores of 30 or more.

If you are curious to know where the answer 15 comes from, and are confident with probability, read on.

When you roll the dice, there are two possible outcomes.

a) You score points (between 1 and 5, so the average score will be 3) and you will add these points to your total. There is a $\frac{5}{6}$ chance of this happening.
b) You will roll a six and lose your points. There is a $\frac{1}{6}$ chance of this happening.

As you roll the dice, you expected to gain $3 \times (\frac{5}{6}) = \frac{15}{6} = 2\frac{1}{2}$ points, and you expect to lose (all your current points) $\times \frac{1}{6}$. On a score of 15 'on average' your expected loss is $\frac{15}{6}$, or $2\frac{1}{2}$ points – exactly the same as the number of points you expect to gain. So this is the break-even point. On scores of more than 15, the probability says you should expect to lose (on average) more points than you expect to gain, so it's not worth continuing.

PAGE 289

a) The mean weight is $66.2 \div 12 = 5.5$kg (more exactly it is 5.517kg, but it's misleading to quote the answer to more decimal places than were given in the original data).

The median weight is between 5.4 and 5.5kg, so the convention is to pick the midpoint of 5.45kg, which is very close to the mean.
b) Using the first digit of the weight as the stem, we get:

Stem	Leaf
4	7 8
5	0 1 2 4 5 6 8
6	2 2 7

The table shows that most of the babies fall into the middle band (between 5kg and 5.9kg), with the rest falling evenly above and below that range. If drawn as a graph this forms a fairly symmetrical distribution, with a bulge in the middle 5–5.9kg range. This is known as the Normal Distribution (see page 291).

If all the babies had weighed between 5.0 and 5.9kg then the stem-and-leaf table would have revealed very little, since all the babies would have been on the same stem. When data is too clustered, the value of this technique is limited.

c) The modal weight (i.e., the weight that was recorded most frequently) was 6.2kg. However, this number means very little, 6.2kg is hardly a 'typical' weight and if you were to weigh another baby, the data suggests it would be much more likely to weigh less than 6kg rather than more. For data like this, the mode is not a helpful statistic unless you group the weights into bands.

PAGE 300

The number in the survey depends on how accurate your results need to be, but it will always be true that you need far more in the snake survey than the football survey. The reason is that maybe as many as 50% of the public sometimes watch football on TV, whereas most sources would suggest that fewer than one in a hundred keep a snake. So if you were to do a survey of 100 shoppers, a significant number would claim to have watched football, whereas it's quite likely that none would be a snake-owner. Only after interviewing thousands of people would you begin to get a snake statistic of any reliability, be that 1 in 100 or 1 in 1000.

ANSWERS TO PROBLEMS AND PUZZLES

1) THE GAZELLE PUZZLE

a)

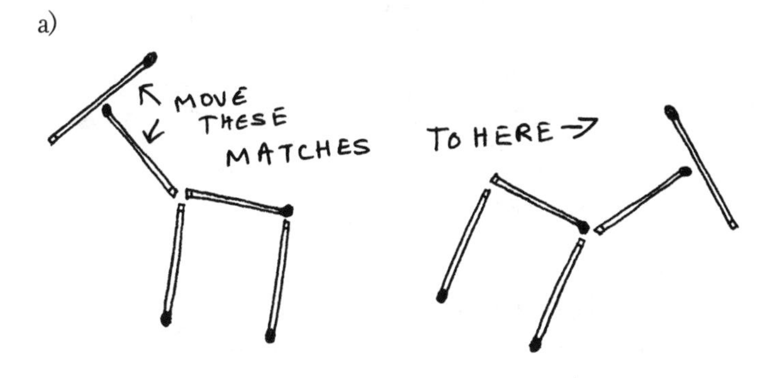

b) It is fascinating to watch people attempting to solve this puzzle, especially if they are sat around a table looking at it from different perspectives. A person looking at the gazelle 'standing on its legs' finds it hard to imagine those matchsticks being anything but legs. Many teenagers and adults attempt to solve it just by staring at it, yet the solution usually comes much faster if somebody 'has a go' and picks up a matchstick, even if they then put it in the wrong place. It is often another person who then suddenly has the 'Aha' moment and spots where to place the matchstick that was removed. It's not unknown for this puzzle to be solved inside thirty seconds, and yet we also know of a very bright physicist who stared at it for over an hour, then thought about it overnight, and finally had to be told the answer. Significantly, that physicist didn't touch the matchsticks once, thus breaking at least one golden rule of problem-solving which is: *always try things out*.

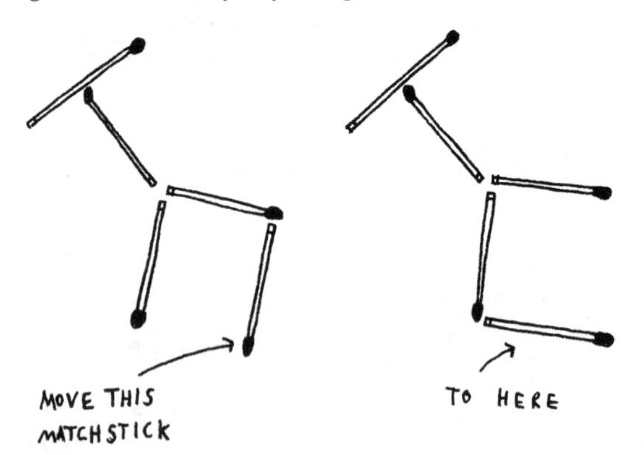

2) CLOCK HANDS

The total number of times that the clock hands are at right angles between noon and midnight is twenty-two. A lot of teenagers (and adults) get stuck at two: they realise 3 p.m. and 9 p.m. but then, since there aren't any other 'o'clocks' that work, they stop. Many others will erroneously think that the hands are at right angles at 3:30 p.m., not stopping to realise that while the minute hand is pointing down at half past, the hour hand is midway between 3 and 4. However, there is a time, about 3:33, when the hands do form 90°. It isn't necessary to work out exactly what that time is, just to realise that the same thing happens at just after 12:15 p.m., then before 12:50 p.m., at about 1:20 p.m. and 1:55 p.m., and so on throughout the twelve hours. However, this is where it's important to be systematic, because otherwise another common mistake is to guess that it happens once every hour when the minute hand is ahead of the hour hand, and the same when the minute hand is behind the hour hand, making twenty-four times in total. This is not quite the case as the table below shows:

The approximate times when the hands are at 90°:

12:16	12:50
1:22	1:55
2:27	3:00
3:33	4:05
4:38	5:11
5:44	6:16
6:49	7:22
7:55	8:27
9:00	9:32
10:05	10:38
11:11	11:44

Count them up and there are only twenty-two occasions between noon and midnight when the hands are at right angles. Where are the 'missing' two? Look down the hours and you'll see that the next time after 2:27 is 3:00, there is no 'Two fifty-something', after 8:27 the next time is 9:00, there is no 'Eight fifty-something'.

This sort of systematic thinking to avoid double counting is used by everyone from scientists to accountants, and is only learned through hard experience. Maths problems can be good training for it.

3) THE DICTIONARIES

The two volumes of the dictionary are each 5cm thick, but the bookmarks are separated by less than 1cm: they are practically next to each other. Why? Because when a book is on the shelf, the first page of that book is on the right-hand side as you look at the spine, not the left. Just imagine taking the book off the shelf and opening it (or, if that doesn't convince you, grab a physical book off the shelf). Most people get this puzzle wrong, and suffer a bit of a 'Doh!' moment. In a way, it serves as little more than a 'trick' question unless you warn the person solving it to check their first answer as it's likely to be wrong – in other words, you are turning an apparently trivial question into a problem. This is an example of when drawing a quick sketch can help.

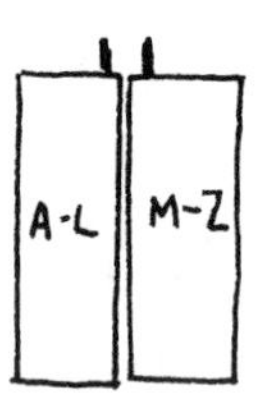

4) THE BABYSITTER

We've included this problem as a reminder that real life rarely presents problems where maths is the only issue. If this were purely a maths problem, then it might make sense to say: '£20 in three hours means the rate is $\frac{20}{6}$ pounds per half-hour, which means $\frac{100}{6}$ = £16.67 for two and a half hours.' Refreshingly, though, most teenagers recognise that there's more to it than this, and the answers that they put forward in discussion involve various other factors. On the one hand, you might be expected to pay your bus fare as part of the fee, so even if you could only babysit for five minutes there might be a cost. On the other hand, what about the inconvenience to the people you're sitting for? Maybe they'll be late for the show they are attending, maybe they'll even have to cancel? So we had teenagers offering answers such as 'Round it down to £15' and even 'Don't charge anything, since you've failed to deliver to your agreement.' There's hope for the future of British customer service yet!

5) DATE PUZZLE

The cubes need to be able to make every number from 01 to 31. This includes 11 and 22, so there must be a 1 and a 2 on each cube. And what about 01, 02, 03, 04 and so on? There will have to be a 0 on each cube as well.

First cube	Second cube
0	0
1	1
2	2
?	?
?	?
?	?

This leaves 3, 4, 5, 6, 7, 8 and 9 still to put on to the blank faces of the cube. There are seven numbers to fit on, yet there are only six blank faces. It can't be done! In fact, you have now solved the problem: it's not possible to squeeze all ten digits onto the two cubes to make all of the dates. At least, not unless you have one final 'Aha' moment – when you realise that to make a 9, all you need to do is turn a 6 upside down. (Lest you think this is a trick, there are cube calendars on the market which do exploit the upside-down 6 in order to work.)

Here is one combination that works: first cube 0, 1, 2, 3, 4, 5, second cube 0, 1, 2, 6, 7, 8.

6) TWO PLUGHOLES

If the Bow and Arrow question (page 5 for a reminder) causes teenagers angst for not being as easy as it looks, then the plughole question leaves them absolutely floored. The most common first guess is that since the plugholes empty the bath in four minutes and six minutes on their own, then the answer when both are open must be the average, five minutes. Yet this can't be true, since it would mean having both plugs out empties the bath more slowly than having only one plug out. The next guess is to assume the answer will be to take the difference: six minutes minus four minutes = two minutes. That answer sounds plausible, and in fact it isn't too far from the right answer, but the reasoning doesn't stand up to testing. What if both plugs empty the bath at the same rate, for example, if each of them takes six minutes? If the formula is simply to work out the difference between the two times then when both plugs are taken out, this would mean that it takes 6 - 6 = ZERO minutes to empty the bath!

Remember how we said that one of the main challenges in solving maths problems isn't the maths itself, it's working out what the problem actually is. That's certainly the case here. This is a question about the rate at which something happens, and these types of question are notoriously difficult to get your head around. Indeed, you shouldn't worry if you or your teenager find the plughole question hard, since even the majority of first-year maths undergraduates struggle to answer it correctly (which we do find rather worrying).

And yet . . . when the problem is set out properly, the answer is very easy to find.

The first plughole empties in six minutes, in other words it empties $\frac{1}{6}$ of a bath per minute. The second plughole empties $\frac{1}{4}$ bath per minute. So after one minute, the total amount of bath that has been emptied is $\frac{1}{6} + \frac{1}{4} = \frac{5}{12}$. So with both plugs out, water leaves the bath at $\frac{5}{12}$ of a bath per minute. So if it takes one minute to empty $\frac{5}{12}$ of a bath, then it will take $\frac{12}{5}$ minutes to empty a whole bath (this is a classic application of ratios, see page 48). $\frac{1}{5}$ of a minute is 12 seconds, so the bath empties in 144 seconds, or 2 minutes 24 seconds.

7) HOW MANY SQUARES?

This is another problem that lends itself to methodical thinking. In this case it also helps to start by looking at a simple problem. Instead of a complicated 8×8 chessboard, think about a 2×2 chessboard:

There's one outer square, and four small squares, so $1 + 4 = 5$ in total.

How about a 3×3 square?

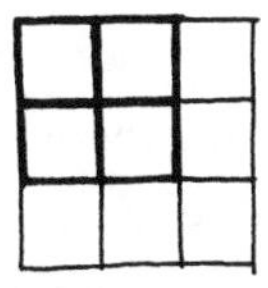

There's one large outer square, and nine small squares. But there are also four 2×2 squares like the one indicated in bold. That makes $1 + 4 + 9 = 14$ in total.

As you increase the size of the chessboard, it becomes clear that the total number of squares in the chessboard is always the sum of the 'square numbers' (1, 4, 9, 16), so the number of squares in an 8×8 chessboard is $1 + 4 + 9 + 16 + 25 + 36 + 49 + 64 = 204$. There's no easy shortcut to that answer, but now that you know the pattern, with a spreadsheet you could easily work out the number of squares in a 100×100 chessboard in a matter of seconds.

8) ESCAPING PRISONERS

Mathematicians are notorious for coming up with an instant answer to this question – and then discovering that their answer is wrong! Their common instant reaction is: 'It's going to be the prime numbers who escape, isn't it!' (Answer: NO, it's not the prime numbers.)

Those with less maths knowledge, on the other hand, might begin to speculate that maybe it's the odd numbers who escape, or that everyone will escape or . . . well, the only way to find out is by solving the problem.

This is another example where simplifying the problem and then being methodical pays off. Instead of a hundred cells, start with ten.

The first prison guard visits and unlocks every cell. The second guard then comes round and turns the key in all of the even-numbered cells, hence locking them. The situation can be set out in a table (× means locked, ✓ means open). The table opposite shows what happens after the first ten guards visit the cells.

After the tenth prison guard leaves, cells 1, 4 and 9 are unlocked, while the rest are locked. And since every prison guard from now on is numbered larger than 10, these cells won't get any more visitors, so the prisoners in 1, 4 and 9 will escape and the rest, 2, 3, 5, 6, 7,8 and 10, will not. The numbers 1, 4

PRISON CELL NO: AFTER VISIT BY PRISON OFFICER NO...	1	2	3	4	5	6	7	8	9	10
1	✓	✓	✓	✓	✓	✓	✓	✓	✓	✓
2	✓	✗	✓	✗	✓	✗	✓	✗	✓	✗
3	✓	✗	✗	✗	✓	✓	✓	✗	✗	✗
4	✓	✗	✗	✓	✓	✓	✓	✓	✗	✗
5	✓	✗	✗	✓	✗	✓	✓	✓	✗	✓
6	✓	✗	✗	✓	✗	✗	✓	✓	✗	✓
7	✓	✗	✗	✓	✗	✗	✗	✓	✗	✓
8	✓	✗	✗	✓	✗	✗	✗	✗	✗	✓
9	✓	✗	✗	✓	✗	✗	✗	✗	✓	✓
10	✓	✗	✗	✓	✗	✗	✗	✗	✓	✗

and 9 should ring a bell. They are the square numbers and in fact it is all the 'square' prisoners, 1, 4, 9, 16, 25, 36, 49, 64, 81 and 100, who escape.

In case you want to know *why* the square numbers escape (and it's understandable if you don't), it's all to do with factors, a topic discussed further on page 97. Each prison cell is visited by the prison guards whose number divides exactly into that cell number. These are called the cell number's factors. For example, the factors of 10 are 1, 2, 5 and 10, and it is prison guards 1, 2, 5 and 10 who visit cell number 10. These factors can be paired up, so $10 = 1 \times 10 = 2 \times 5$, and in fact all numbers have an even number of factors (so an even number of prison guards visits) except for square numbers. The factors of 9, for example are 1×9 and 3, while the factors of 36 are 1×36, 2×18, 3×12, 4×9 and a solitary 6. An even number of visits means a cell ends up locked while an odd number means it ends up unlocked. So it is the numbers that have an odd number of factors – the squares – who escape.

9) THE DIE-HARD JUGS PUZZLE

This is something of a classic, but nice for the fact that experimentation usually leads to a solution (and there are other possible solutions, some more efficient than others).

The solution used in the film goes like this:

a) Fill the 5-litre jug
b) Use the 5-litre jug to fill the 3-litre jug, leaving 2 litres.
c) Empty the 3-litre jug and pour the 2 litres from the 5-litre jug into it.
d) Refill the 5-litre jug.
e) Top up the 3-litre jug with one more litre from the 5-litre jug, leaving 4 litres.

10) THE STATUE IN THE DESERT

This problem resembles real life more than any of the others because it doesn't give you any clues as to which information is relevant to solving the problem. Life is full of redundant information and a crucial part of problem-solving is being able to work out what is useful from what isn't. In this case there are several possible solutions. The one that many teenagers leap to first is to attempt to use the phone to ring a friend and request a very long tape measure and a ladder. Unfortunately this solution is likely to fail at the first hurdle: the chance that there will be any phone reception in the middle of the desert is almost zero.

One of the most elegant solutions that would actually work is to use the idea of similar triangles. Wait until the sun comes out, then use the string to mark out the length of the shadow of the statue from its base to its tip and then measure the length of the string using the tape measure. Then stand up the water bottle (say) and measure the length of its shadow and its height. The ratio of the height of the water bottle to its shadow length is the same as the ratio of the height of the statue to its shadow, hence the statue height can be calculated.

Another possibility is to improvise a crude protractor using the writing pad. All you need on the protractor is the angle of 45°, which you can make by taking a rectangular piece of paper and folding one corner up to meet the side of the rectangle:

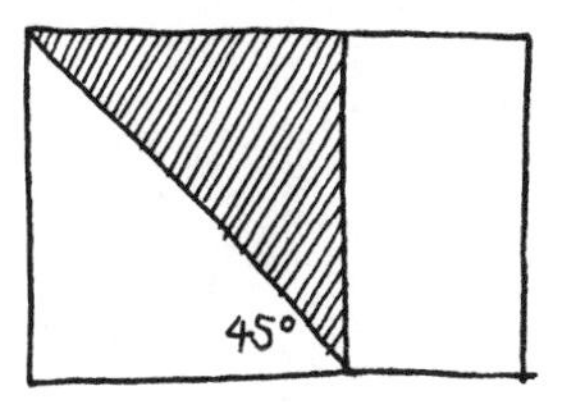

Using the mobile phone as a weight, hanging from a piece of string held to the corner of the protractor, tilt up the protractor so the string lines up with the diagonal. Now walk to the position where your eye lines up along the edge of the writing pad up to the top of the statue. You have now formed an isosceles triangle – there are two angles of 45°, so the height of the statue above your eye level must be the same as the distance from your feet to the base of the statue (see the diagram). Now all you have to do is

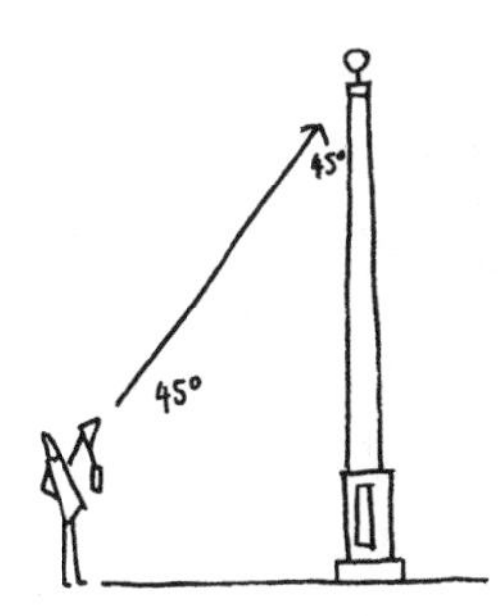

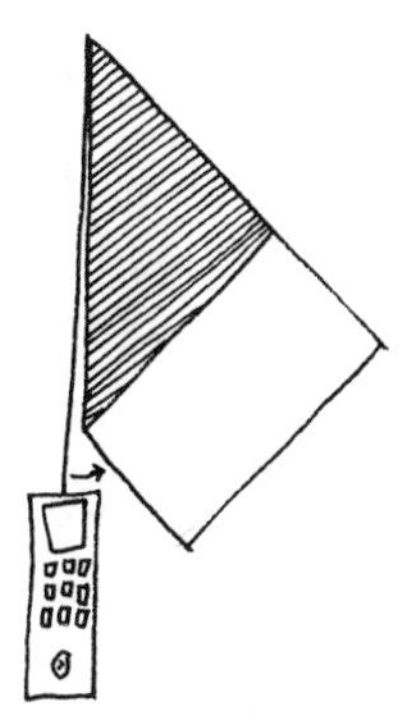

measure the distance from your feet to the base of the statue – and then add on the height of your eyes above the ground!

Then there are the more wacky solutions. For example, find a local tribesman, and tell him that if he tells you how high the statue is, you'll give him a free pair of binoculars.

GLOSSARY

This glossary contains common mathematical terms that you might encounter, some of them dealt with in more detail in the rest of the book.

Acute angle Angle that is less than 90°.

Algorithm Rather like a recipe, an algorithm is a set of mathematical instructions to be carried out to make a particular calculation.

Angle notation An angle will usually be indicated with a small letter (greek letters like theta, θ, are popular for this, but you'll find 'x' or 'a' almost as often). But if the corners of a shape are labeled, then the angle between two sides of the shape is often indicated using three letters. The middle letter refers to the corner whose angle you want to know, the other two letters are the two corners that it is joined to.

Arc Part of the perimeter of a circle.

Associative law The associative law applies to addition and multiplication, and can be summed up as: 'Brackets make no difference.' For example, when doing addition, adding $3 + 8 + 2$ produces the same answer (13) regardless of whether you first add $(3 + 8)$ or $(8 + 2)$. The same is true of multiplication, where $(8 \times 12) \times 2 = 192$, as does $8 \times (12 \times 2) = 192$. The law doesn't apply to subtraction or division. For example, the subtraction $10 - 5 - 2$ depends on whether you calculate it as: $(10 - 5) - 2 = 3$, or $10 - (5 - 2) = 7$. See also 'Commutative law'.

Axis Most often an axis means the reference lines (x and y axes) from which distances are measured in the co-ordinate system. It can also refer to a line through which a shape is reflected, or around which an object is rotated.

Base The building block of any number system. Usually we count in base ten (because that is the number of fingers we have). Counting in base five (for example), the numbering would be: 1, 2, 3, 4, 10, 11, 12, 13, 14, 20 etc. The number 10 here means one base number (5) and zero units.

Bearings Traditionally used in navigation (by ships or by people walking up hills), bearings are the angle between the direction you are aiming and north, measured clockwise in degrees. 90° represents east, 315 degrees is north-west.

Bisector A line that divides an angle, line or shape in half.

Chord A straight line across a circle.

Circle theorem Any general rule connecting the angles and lines within a circle. Although circle theorems can get complicated, they are all ultimately derived from the basic proofs about angles, circles and triangles that are discussed in the geometry chapters.

Coefficient In the expression $6x$, 'x' is the variable and 6 is the coefficient, so the coefficient is the number that multiplies a variable in an expression.

Combinations The number of different selections that can be made from a group of items. For example, the number of two-letter combinations that can be made from the letters A, B and C (without any repetition) is three: AB, AC and BC. See also Permutations.

Completing the square Technique used to solve quadratic equations.

Commutative law It doesn't matter in which order you add or multiply two numbers: $3 + 7 = 7 + 3$, and $14 \times 4 = 4 \times 14$. The law doesn't apply to subtraction or division ($8 - 3 \neq 3 - 8$).

Congruence Two shapes are congruent if one can be placed over the other and be a perfect fit. If one of the shapes needs to be 'flipped over' to cover the other that still counts. The main use of congruence is as a precise way of conveying that two shapes in different parts of a diagram are the same. Usually this will be in a geometrical proof, along the lines of: 'Shape ABCD is congruent with shape WXYZ and hence . . .'

Cube root This is the reverse of the cube of a number, 4^3 (four 'cubed') $= 64$, the cube root of 64 is 4.

Cuboid A cuboid is any solid shape with six rectangular faces – a regular cardboard box, for example.

Cumulative frequency In a spread of data, the cumulative frequency is a running total of the occurrences so far, starting from the first (usually the smallest) data.

Denominator The bottom number of a fraction.

Dimensions Usually the length, breadth and height of an object.

Distribution How the data is spread out; for example, the data could be evenly spread or it might be bunched up in the middle forming a peak or a bell shape.

Distributive law When multiplying brackets that contain an addition or a subtraction, you get the same answer whether you multiply each number in the brackets separately, or do the calculation in brackets first. For example, $7 \times (3 + 4) = 7 \times 3 + 7 \times 4 = 7 \times 7 = 49$.

Equation Any mathematical statement that contains an equals sign is an equation, for example, $10 = 3 + 7$ or $2x = 3y + 7$

Equivalent fractions Fractions with the same value as each other, e.g. $\frac{1}{2}$ and $\frac{3}{6}$

Expression Any mathematical statement that contains numbers, variables (such as x or y) and operators (such as + or ×)

Exponential A number or variable raised to a power, such as 2^5 or 7^x. In an exponential series such as 3, 6, 12, 24, each term is a fixed multiple of the previous term (in this example, each term is double the previous one).

Factors Numbers or expressions that divide exactly into other numbers or expressions with no remainder. For example 7 is a factor of 21, and x is a factor of $3x^2$.

Gradient The steepness of a line, worked out by dividing the distance up by the distance across.

Highest common factor (or HCF) If you have two numbers, the HCF is the largest whole number that will divide into both of them. The HCF of 28 and 42 is 14.

Hypotenuse The longest side of a right-angle triangle.

Integer Any whole number, including negative numbers and zero.

Inverse operations Two operations are inverse of each other if one 'undoes' the other; for example, squaring and taking the square root are inverse operations.

Inverse relationship Two variables have an inverse relationship if increasing one decreases the other. An example is temperature and the coverage of snow – as temperature increases, snow cover decreases.

Irrational number A number that can not be expressed as the ratio of two whole numbers. Pi and the square root of 2 are irrational numbers.

Isosceles A triangle with two sides that are the same length.

Linear equation An equation which when plotted as a graph forms a straight line. For example $y = 2x + 3$.

Logarithm Logarithms are linked to the idea of numbers being raised to powers, for example 2^3 ('2 to the power 3', $= 2 \times 2 \times 2$). When you multiply two such numbers together, the answer can be found by adding the raised numbers (powers) together, for example $4^5 \times 4^{11} = 4^{16}$. The powers in this example are called *logarithms in base 4*.

Lowest common denominator (LCD) When comparing two fractions, the lowest common denominator is the smallest number that the denominators (i.e. bottom numbers) of both fractions will divide into. For example with $\frac{3}{10}$ and $\frac{1}{4}$ the LCD is 20, because 20 is the smallest number that divides by 10 and 4. The fractions can now be compared, as $\frac{3}{10} = \frac{6}{20}$ and $\frac{1}{4} = \frac{5}{20}$

Lowest common multiple (LCM) If you have two numbers, their lowest common multiple is the smallest number into which both numbers divide exactly. For example, the LCM of 12 and 18 is 36.

Milli One thousandth of (millimetre, millisecond, etc.)

Nets A flat sheet that can be folded to make a 3D object.

Numerator The top number of a fraction.

Obtuse angle An angle larger than 90° and smaller than 180°.

Operation A mathematical process that changes numbers. The most common are addition, subtraction, multiplication and division.

Parallelogram A four-sided shape in which the opposite sides are all parallel to each other.

Permutations The total number of combinations that can be made from a particular group of items when the order matters. From the letters ABC the number of two-letter permutations is six: AB, AC, BA, BC, CA and CB. See also Combinations.

Perpendicular At right angles to, usually referring to a line that is at right angles to another line.

Polar co-ordinates An alternative way to indicate the position on a map, polar co-ordinates indicate the distance and the angle from 'north' (or whichever direction is chosen as zero).

Polygon Any shape whose sides are all straight lines. Poly means many, gon refers to the angles (derived from the Greek work for knee – the polygon is a shape with many knees).

Product The result when numbers or variables are multiplied together. The product of 4 and 6 is 24.

Quadratic equation An equation which involves a variable being squared, for example $t^2 + 4t + 5 = 0$.

Quadrilateral A shape with four straight sides (quad meaning four, lateral meaning 'sides'). There are many special types of quadrilateral, including the square, rectangle, rhombus, parallelogram, trapezium and kite.

Radians An alternative way of measuring angles, useful in some areas of more advanced maths where they can make calculations considerably easier. One radian is roughly 57°, There are exactly 2π radians in a circle.

Rational number Any number that can be expressed as one whole number divided by another whole number; $\frac{1}{4}$ is a rational number. This includes intergers (since they can always be divided by 1).

Reciprocal The reciprocal of a number is 1 divided by the number. The reciprocal of 8 is $\frac{1}{8}$.

Reflex angle An angle between 180° and 360°.

Regular (shape) A regular shape, such as hexagon or pentagon, is one whose sides and angles are all the same. Note that a shape such as a hexagon might be symmetrical, but still not regular.

Scale factor The amount by which lengths in a shape are magnified or reduced. A square whose sides are doubled in length is magnified by a scale factor of 2.

Scalene triangle A triangle whose sides all have different lengths.

Segment A portion of a circle, the shape of a slice of cake.

Sequence A list of numbers in which each term can be worked out from the previous terms (for example, in a Fibonacci sequence, each term is the sum of the previous two terms).

Series Similar to a sequence, but each term is calculated from a formula based on its position.

Sign A plus or minus sign, indicating whether a number is positive or negative.

Similarity Two shapes are said to be 'similar' if you could enlarge one of them using a photocopier and make it fit perfectly over the other. In the case of triangles, two triangles will be similar if their three angles are identical, or if the ratio of the lengths of the three sides is the same in both. For example, a triangle with sides of length 3, 4 and 5cm is similar to one with sides of 6, 8 and 10cm.

Surd A surd is a number that contains a square root (or some other root, e.g. a cube root), and which cannot be simplified. For example $(1 + \sqrt{2})$ is a surd. However, $\sqrt{9}$ is not, because it can be simplified : $\sqrt{9} = 3$.

Tangent

(1) A straight line that touches a circle at a point. A tangent is always perpendicular to (i.e., forms a right-angle with) the circle's radius:

(2) Tangent (or tan) is the length of the opposite side of a triangle divided by the adjacent side.

Theorem Any mathematical statement that has been proved to be true. All theorems are built up from other theorems. In the case of geometry, you can trace most theorems all the way back to the original theorems stated by Euclid over 2000 years ago.

Translation Sliding a shape to another position without rotating it.

Trapezium A four-sided shape with exactly two sides that are parallel.

Tree diagrams Resembling a tree, this is a way of presenting all the different possible outcomes; for example, the outcomes of tossing a coin. The tree branches every time there is more than one possible outcome.

Triangle number The first triangle number is 1, the second is 1 + 2 (=3), the third is 1 +2 +3 (=6) and so on. You'll find the fifth triangular number 1 +2 +3 +4 +5 (=15) when you look at the balls laid out in a triangle at the start of a game of pool or snooker.

Velocity In everyday life 'velocity' and 'speed' are used interchangeably, but in maths and the sciences velocity means both speed and direction, and is often represented as a vector. Two cars travelling in opposite directions at 50mph have the same speed, but different velocities (one is the negative of the other).

X-axis The horizontal axis used when plotting a graph.

Y-axis The vertical axis used when plotting a graph.

Y-intercept The point at which a graph crosses the y-axis.

ACKNOWLEDGEMENTS

We are indebted to several teachers for their insights and for picking through the early drafts of this book in particular Rachael Horsman and Julian Gilbey who gave us hours of their time despite their busy schedules. Special thanks also to Andrew Hemmings and Karen Collins and their pupils at Darrick Wood Academy, and to the staff and pupils at Watford Girls Grammar, Claydon High School and Simon Langton Girls School, Luke Kerr and students from Mornington High School and Rob and Bea Carter for taking tests.

Many parents and teenagers shared their experiences with us, including Tim and Sophie Baxter, Dan Bevan, Jessica Cargill-Thomson, Fedelma, Tony, Amy and Thomas Good, Fiona and Chris Hewitt, Matthew Page, Rachel Parkinson, Matt Phillips, Kerry Muraszko, David Rogerson, Chris Roles, Fiona and Ian Sweetenham, Ruth Warburton, Claire Wilshaw and Fiona Woods. And thank you Andrew Jeffrey for allowing us to use the quote about the tram timetables.

We were greatly helped by the maths and curriculum expertise of Richard Lissaman, Ali Gross, Kevin Houston, Chris Olley, David Wells, David Acheson and Jason Wanner. The responsibility for any errors that remain in the book lies squarely with the authors.

Finally, thanks to Elaine Standish and Chris Healey for your editing skills, our copy-editor Alison Tulett and proof-reader Jane Howard, Peter Ward for his careful and stylish type design, Rowena Skelton-Wallace, Simon Rhodes, Kris Potter, Lee-Anne Williams, Tom Drake-Lee and the team at Vintage Books and most of all to our wonderfully patient editor Rosemary Davidson for sticking with us through this long project.